SEAWATER: ITS COMPOSITION, PROPERTIES AND BEHAVIOUR

THE OCEANOGRAPHY COURSE TEAM

Authors
Joan Brown
Angela Colling
Dave Park
John Phillips
Dave Rothery
John Wright

Editor
Gerry Bearman

Design and Illustration
Sue Dobson
Ray Munns
Ros Porter
Jane Sheppard

This Volume forms part of an Open University course. For general availability of all the Volumes in the Oceanography Series, please contact your regular supplier, or in case of difficulty the appropriate Pergamon office.

Further information on Open University courses may be obtained from: The Admissions Office, The Open University, P.O. Box 48, Walton Hall, Milton Keynes, MK7 6AA.

Cover illustration: Satellite photograph showing distribution of phytoplankton pigments in the North Atlantic off the US coast in the region of the Gulf Stream and the Labrador Current. *(NASA and O. Brown and R. Evans, University of Miami.)*

SEAWATER: ITS COMPOSITION, PROPERTIES AND BEHAVIOUR

PREPARED BY AN OPEN UNIVERSITY COURSE TEAM

PERGAMON PRESS
OXFORD · NEW YORK · BEIJING · FRANKFURT · SÃO PAULO · SYDNEY
TOKYO · TORONTO

in association with

THE OPEN UNIVERSITY
WALTON HALL, MILTON KEYNES MK7 6AA, ENGLAND

U.K.	Pergamon Press plc, Headington Hill Hall, Oxford OX3 0BW, England
U.S.A.	Pergamon Press, Inc., Maxwell House, Fairview Park, Elmsford, New York 10523, U.S.A.
PEOPLE'S REPUBLIC OF CHINA	Pergamon Press, Room 4037, Qianmen Hotel, Beijing, People's Republic of China
FEDERAL REPUBLIC OF GERMANY	Pergamon Press GmbH, Hammerweg 6, D-6242 Kronberg, Federal Republic of Germany
BRAZIL	Pergamon Editora Ltda, Rua Eça de Queiros, 346, CEP 04011, Paraiso, São Paulo, Brazil
AUSTRALIA	Pergamon Press Australia Pty Ltd., P.O. Box 544, Potts Point, N.S.W. 2011, Australia
JAPAN	Pergamon Press, 5th Floor, Matsuoka Central Building, 1-7-1 Nishishinjuku, Shinjuku-ku, Tokyo 160, Japan
CANADA	Pergamon Press Canada Ltd., Suite No. 271, 253 College Street, Toronto, Ontario, Canada M5T 1R5

First edition 1989

Library of Congress Cataloging in Publication Data
Seawater: its composition, properties, and behaviour
1. Seawater—Composition. I. Open University.
GC101.2.S4 1988 551.46′01—dc 19 88-19579

British Library Cataloguing in Publication Data
Seawater
1. Seawater. Chemical analysis
I. Open University. *Oceanography Course Team*
551.46′01
ISBN 0-08-036368-7 (Hardcover)
ISBN 0-08-036367-9 (Flexicover)

Jointly published by the Open University, Walton Hall, Milton Keynes, MK7 6AA and Pergamon Press plc, Headington Hill Hall, Oxford OX3 0BW.

Designed by the Graphic Design Group of The Open University.

Printed in Great Britain by A. Wheaton & Co. Ltd, Exeter.

CONTENTS

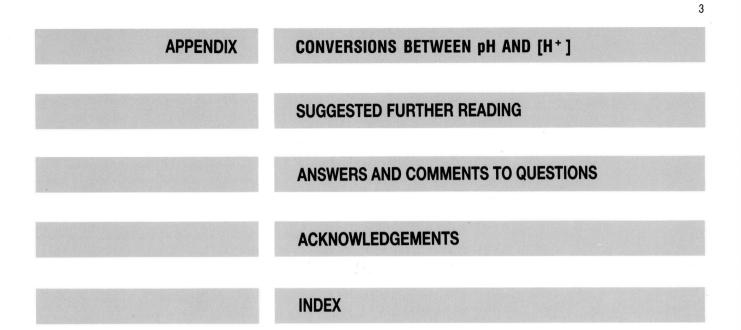

ABOUT THIS VOLUME

This is one of a Series of Volumes on Oceanography. It is designed so that it can be read on its own, like any other textbook, or studied as part of S330 *Oceanography*, a third level course for Open University students. The science of oceanography as a whole is multidisciplinary. However, different aspects fall naturally within the scope of one or other of the major 'traditional' disciplines. Thus, you will get the most out of this Volume if you have some previous experience of studying chemistry and a certain amount of physics. Other Volumes in this Series lie variously within the fields of geology, biology, physics or chemistry.

Chapter 1 summarizes the special properties of water and the role of the oceans in the hydrological cycle. Chapters 2 to 4 discuss the distribution of temperature and salinity in the oceans and their combined influence on density, stability and vertical water movements. Chapter 5 describes the behaviour of light and sound in seawater and provides examples of the application of acoustics to oceanography. Chapter 6 examines the composition and behaviour of the dissolved constituents of seawater, covering both minor and trace constituents and the major ions, as well as dissolved gases and biologically important nutrients. It deals also with such topics as residence times, speciation and redox relationships.

Finally, Chapter 7 provides a short review of ideas about the history of seawater, the involvement of the oceans in global cycles and their relationship to climatic change, with special reference to the carbon dioxide problem.

You will find questions designed to help you to develop arguments and/or test your own understanding as you read, with answers provided at the back of this Volume. Important technical terms are printed in **bold** type where they are first introduced or defined.

CHAPTER 1 WATER, AIR AND ICE

All the water and air now at the Earth's surface used to be inside the Earth. It has been progressively released from the Earth's interior by a process of de-gassing that has been going on since the Earth formed about 4.6 billion years ago. The rate of de-gassing has decreased through time because the radioactive elements responsible for much of the Earth's internal heat have been decaying exponentially—so there is much less of these elements now than there was when the Earth formed. In short, the interior of the early Earth was hotter than it is now; convection in the Earth's mantle was more vigorous; and de-gassing was more rapid. It seems likely that most of the oceans and atmosphere had been de-gassed from the Earth's interior by about 2.5 billion years ago, and that de-gassing has continued ever since but at a progressively decreasing rate. Small amounts of water and atmospheric gases continue to be expelled from inside the Earth even today.

The oceans and atmosphere together provide our fluid environment. The nature of that environment is controlled to a very large extent by the special properties of a substance we take virtually for granted: water.

1.1 THE SPECIAL PROPERTIES OF WATER

> 'From a drop of water, a logician could infer the possibility of an Atlantic or a Niagara, without having seen or heard of one or the other.'
>
> Sherlock Holmes, in *A Study in Scarlet*, by Sir Arthur Conan Doyle.

It is easy enough, perhaps, to infer the *existence* of oceans from a drop of water, less easy to deduce that they have waves, tides and currents, still less easy to predict patterns of water movement and water chemistry, and the nature of marine life forms. Nonetheless, a knowledge of the properties of water does enable us to understand at least some of the major characteristics of the oceanic environment.

QUESTION 1.1 Most people know that the oceans are salty, cold, dark and teem with noisy life, and that they are never still. Explain these characteristics of the oceans by selecting items from the following list of properties and attributes of water.

Water is a highly mobile liquid
Water is a good solvent
Water is a poor conductor of heat
Water has a high specific heat
Water has a high latent heat of fusion and of evaporation
Pure water freezes at 0°C
Pure water boils at 100°C
The maximum density of freshwater is at 4°C; for seawater it is at its freezing point (−1.9°C)
Ice is less dense than water
Light can only travel a maximum of a few hundred metres through water
Sound can travel thousands of kilometres through water
Water is essential to life

Table 1.1 Anomalous physical properties of liquid water.

Property	Comparison with other substances	Importance in physical/biological environment
specific heat ($=4.18\times10^3$ J kg^{-1}°C^{-1})	highest of all solids and liquids except liquid NH_3	prevents extreme ranges in temperature; heat transfer by water movements is very large; tends to maintain uniform body temperatures
latent heat of fusion ($=3.33\times10^5$ J kg^{-1}°C^{-1})	highest except NH_3	thermostatic effect at freezing point due to the absorption or release of latent heat
latent heat of evaporation ($=2.25\times10^6$ J kg^{-1})	highest of all substances	large latent heat of evaporation is extremely important in heat and water transfer within the atmosphere
thermal expansion	temperature of maximum density decreases with increasing salinity; for pure water it is at 4°C	freshwater and dilute seawater have their maximum density at temperatures above the freezing point; the maximum density of normal seawater is at the freezing point
surface tension ($=7.2\times10^9$ N m^{-1})*	highest of all liquids	important in the physiology of the cell; controls certain surface phenomena and the formation and behaviour of drops
dissolving power	in general dissolves more substances and in greater quantities than any other liquid	obvious implications in both physical and biological phenomena
dielectric constant** (=87 at 0°C, 80 at 20°C)	pure water has the highest of all liquids except H_2O_2 and HCN	of utmost importance in the behaviour of inorganic dissolved substances because of the resulting high dissociation
electrolytic dissociation	very small	a neutral substance, yet contains both H^+ and OH^- ions
transparency	relatively great	absorption of radiant energy is large in infrared and ultraviolet; in the visible portion of the energy spectrum there is relatively little selective absorption, hence pure water is 'colourless' in small amounts; characteristic absorption important in physical and biological phenomena
conduction of heat	highest of all liquids	although important on a small scale, as in living cells, the molecular processes are far outweighed by turbulent diffusion
molecular viscosity ($=10^{-3}$ N s m^{-2})*	less than most other liquids at comparable temperature	flows readily to equalize pressure differences

*N = Newton = unit of force in kg m s^{-2}.

**Measure of the ability to keep oppositely charged ions in solution apart from one another.

Notes to Table 1.1

1 Latent heat is the amount of heat required to melt unit mass of a substance at the melting point.

2 Specific heat is the amount of heat required to raise the temperature of unit mass of a substance by one degree.

3 Surface tension is a measure of the 'strength' of the liquid surface and hence of the 'durability' of drops and bubbles. (See Section 2.2.1.)

4 Viscosity is a measure of resistance to distortion (i.e. flow) of a fluid. The greater the viscosity, the less readily will the fluid flow (motor oil is more viscous than water).

Table 1.2 Density of pure water at different temperatures.

Temperature (°C)	State	Density (kg m^{-3})
–2	solid	917.2
0	solid	917.0
0	liquid	999.8
4	liquid	1000.0
10	liquid	999.7
25	liquid	997.1

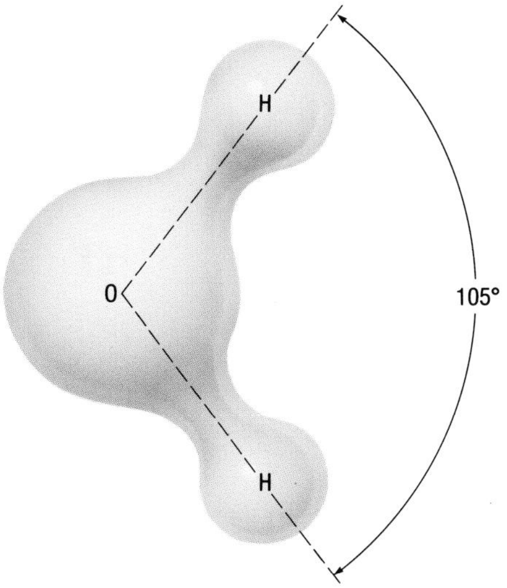

Figure 1.1 Schematic view of the water molecule. It is electrically polarized. The oxygen side carries a small negative charge; the hydrogen side carries a small positive charge.

The molecular mass of water is 18. Comparison with other hydrogen compounds of comparable molecular mass suggests that water should freeze at about −100°C and boil at about −80°C, instead of at 0°C and 100°C respectively (e.g. methane, with a molecular mass of 16, freezes at −183°C and boils at −162°C). The density of most solids is greater than that of their corresponding liquids, and the density of liquids typically decreases progressively when heated from the melting point—but ice is less dense than water, and the maximum density of pure water is reached at 4°C. Tables 1.1 and 1.2 contain much of the same information as that summarized for Question 1.1, but in a more detailed and quantitative form.

The reasons for these anomalous properties of water lie in its molecular structure. A water molecule consists of an oxygen atom bonded to two hydrogen atoms. The angle between the interatomic bonds is 105°. The difference in electronegativity (electrical properties) between oxygen and hydrogen atoms results in the hydrogen side carrying a small positive charge, while the oxygen atom carries a small negative charge (Figure 1.1). Because of this polar structure, water molecules have an attraction for one another and tend to arrange themselves into partly ordered groups, linked by weak intermolecular bonds called hydrogen bonds.

As the temperature of pure water is raised above its melting point (0°C), the energy of the molecules increases, counteracting the tendency to form partly ordered groups. Individual molecules can then fit together more closely, occupying less space and increasing the density of the water. However, raising the temperature also causes thermal expansion of the molecules, which results in decreased density. At temperatures between 0°C and 4°C, the 'ordering effect' predominates, whereas at higher temperatures thermal expansion is more important. The combination of the two effects means that the density of pure water is greatest at 4°C (Table 1.2).

1.1.1 THE EFFECT OF DISSOLVED SALTS

Any substance dissolved in a liquid has the effect of increasing the density of that liquid. The greater the amount dissolved, the greater the effect. Water is no exception. The density of freshwater is close to 1.00×10^3kg m^{-3} (*cf.* Table 1.2), while the average density of seawater is about 1.03×10^3kg m^{-3}.

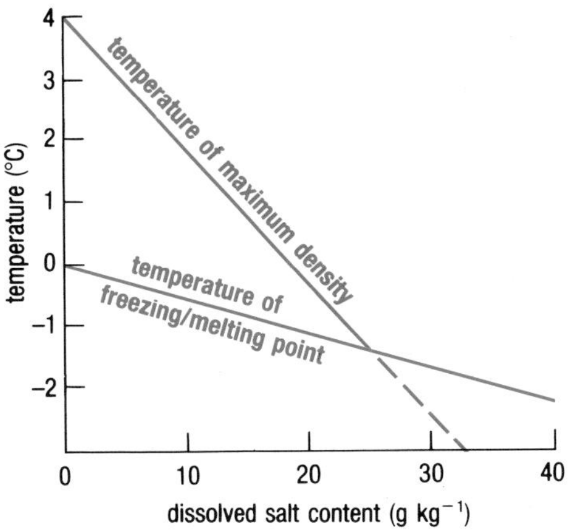

Figure 1.2 Temperatures of freezing and melting point and maximum density of liquid water as functions of dissolved salt content.

Another important effect of dissolved substances is to depress the freezing point of liquids. The addition of salt to water lowers its freezing point (which is why salt is used for spreading on frozen roads). It also lowers the temperature at which water reaches its maximum density, which is 4°C for pure freshwater (Table 1.2). That is because dissolved salts inhibit the tendency of water molecules to form ordered groups, so that density is controlled only by the thermal expansion effect. Figure 1.2 shows that the freezing point and the temperature of maximum density are the same when the salt content (salinity) of water reaches about 25g kg^{-1}. As the oceans are more saline than this (containing on average about 35g kg^{-1} salts), the density of seawater increases with falling temperature right down to the freezing point. This is a crucial distinction between freshwater and seawater and it has a profound effect on oceanic circulation processes and on the formation of sea-ice, as you will see in later Chapters.

QUESTION 1.2 On Figure 1.2, do the words 'maximum density' refer to a single density value, or does the maximum density itself increase or decrease along the line, with falling temperature and increasing salt content?

1.2 THE HYDROLOGICAL CYCLE

The oceans dominate the **hydrological cycle** (Figure 1.3), for they contain 97% of the global water inventory. Large changes in terrestrial parts of the water inventory (Table 1.3) would be necessary to have any significant effect on the amount of water in the oceans. For example, it is estimated that during the glacial maxima of the past two million years, some 50000×10^{15} kg of water were added to the world's glaciers and ice-caps, increasing their volume to about two-and-a-half times what it is today.

Figure 1.3 The hydrological cycle, showing the Earth's water inventory, annual movements of water through the cycle (black numbers), and amounts of water stored in different parts of the cycle (blue numbers). Note the overwhelming dominance of ocean water in the inventory. All quantities shown are $\times 10^{15}$ kg.

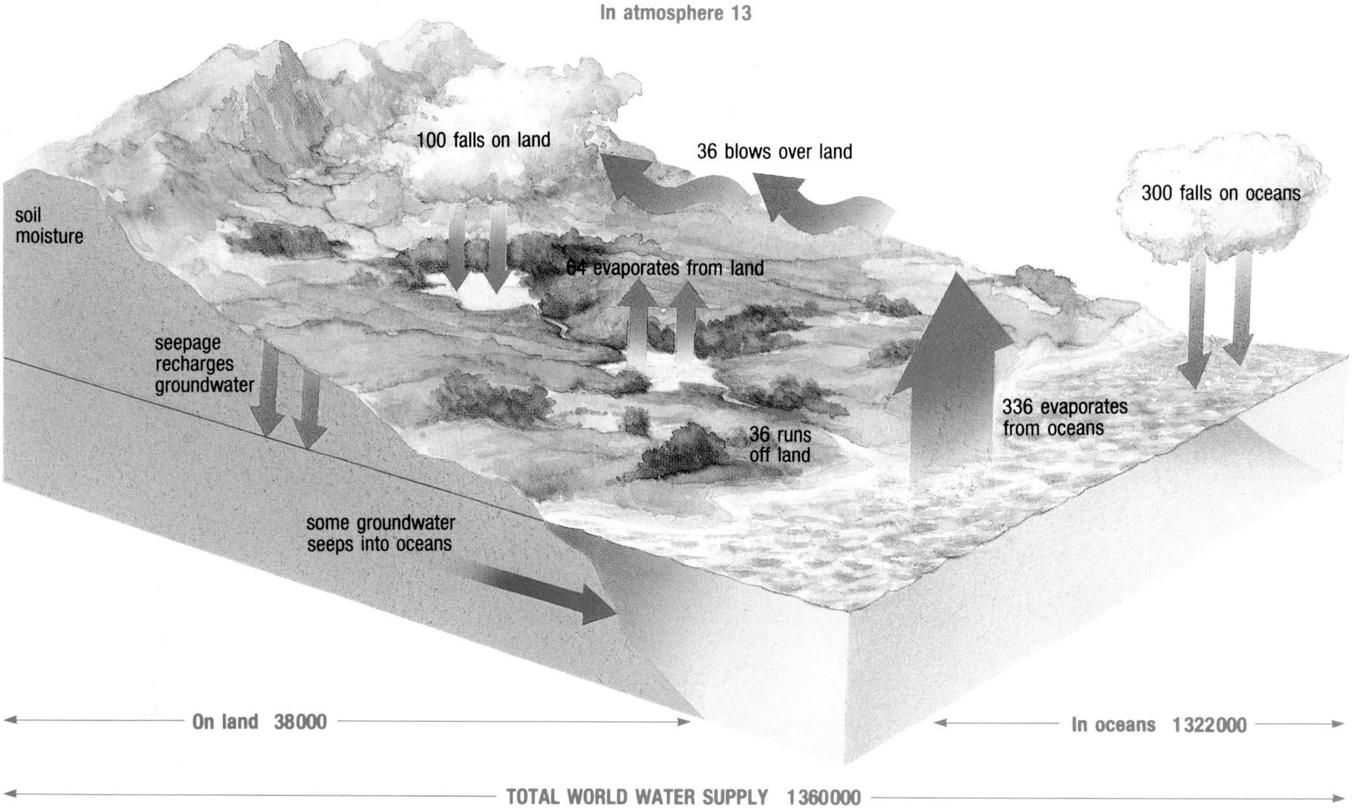

Table 1.3 Water on land

rivers and streams	1
freshwater lakes	125
salt lakes and inland seas	104
total surface water	230
glaciers and ice-caps	29 300
soil moisture and seepage	70
ground water	8400
total on land	38 000

This lowered sea-level world-wide by over 100m—enough to turn most shallow continental-shelf seas into dry land—but it only reduced the total volume of water in the oceans by about 3.5%.

The concept of **residence time** can be defined by reference to Figure 1.3. It is the average length of time that a water molecule resides—or is stored—in any particular stage of the hydrological cycle. It is calculated by dividing the amount of water in that part of the cycle by the amount that enters (and leaves) it in unit time. (You will encounter residence time in other contexts later.)

QUESTION 1.3 (a) What is the annual rate of evaporation from the ocean? Is it balanced by precipitation plus run-off from land?

(b) What is the residence time of water in the oceans?

(c) Approximately what quantity of water moves through the atmosphere annually?

1.2.1 WATER IN THE ATMOSPHERE

The most obvious manifestations of water in the atmosphere are clouds and fog. Both consist of water droplets or ice crystals that have condensed round (or nucleated on) small particles in the air. Water in the atmosphere is mostly in the gaseous state, i.e. as water vapour. Air is saturated with water vapour when there is equilibrium between evaporation and condensation. The higher the temperature, the greater the amount of energy available for evaporation, so warm air can hold more moisture at saturation than cold air.

There are two ways in which unsaturated air can be cooled so that it becomes saturated and condensation begins. Cooling occurs either by adiabatic expansion of the air when it rises (adiabatic processes are discussed in Chapter 4) or when air comes in contact with a cold surface (e.g. condensation on windows). Fogs develop when a sufficiently thick layer of air is cooled in this way, forming in effect clouds at ground (or water) level. Two main types of fog are recognized.

Radiation fog forms when the ground surface is cooled by radiant heat loss at night into a clear sky. If the air in contact with the ground is close to saturation and its temperature falls sufficiently, then fog may form. Radiation fogs do not develop over lakes or the sea, because water has a high specific heat (Table 1.1), so water surfaces cool less rapidly than ground surfaces. However, radiation fog often drifts from land over rivers, estuaries and coastal waters.

Advection fog forms when warm humid air moves (is advected) over cold ground or water and is thus cooled. Such fogs commonly develop over the Grand Banks off Newfoundland for example, where air formerly above the warm Gulf Stream is advected over the cold Labrador Current.

1.2.2 ICE IN THE OCEANS

Polar ice-caps are a significant feature of the present-day Earth. A layer of ice with its covering of snow reflects back more incoming solar radiation than areas of land or open water. Only a small amount of solar energy is transferred to surface waters under the ice, so that ice-layers, once established, tend to be self-perpetuating.

Sea-ice is formed by the freezing of seawater itself. Various stages and ages of sea-ice formation are illustrated in Figure 1.4. When seawater first begins to freeze, relatively pure ice is formed, so that the salt content of the surrounding seawater is increased, which both increases its density and depresses its freezing point further (*cf.* Figure 1.2). Most of the salt in sea-ice is in the form of concentrated brine droplets trapped within the ice as it forms. Brine trapped in sea-ice is much more saline than the ice itself. Its freezing point is greatly depressed relative to that of the ice (*cf.* Figure 1.2); and so the brine droplets remain liquid at temperatures well below those of ice formation.

QUESTION 1.4 (a) Give two reasons why these brine droplets will be denser than the surrounding ice.

(b) Old sea-ice is less saline than young sea-ice formed under comparable conditions, i.e. it contains fewer brine droplets. Can you explain how that happens?

Round the South Pole the centre of ice accumulation is the ice-buried continent of Antarctica, which is surrounded by a shelf of floating ice. In contrast, the North Pole is surrounded by the basin of the Arctic Ocean, which is largely covered by floating sea-ice. In both regions, floating ice-fields are called pack-ice (Figure 1.4).

(a) (b) (c) (d) (e)

Figure 1.4 Forms of sea-ice.
(a) Grease ice or frazil ice, formed as ice crystals grow and coalesce to give the surface an oily appearance.
(b) Pancake ice, formed as grease ice thickens and breaks up into pieces, usually about 0.5–3 m across. The roughly circular shapes and raised rims result from continued collision with one another.
(c) Open pack-ice, formed of ice-floes with many spaces (leads) between them.
(d) Close pack-ice, formed of ice-floes mostly in contact, with few leads.
(e) Brash ice, fragments not more than about 2 m across, the 'wreckage' of other forms of ice, here seen stranded at a shoreline.

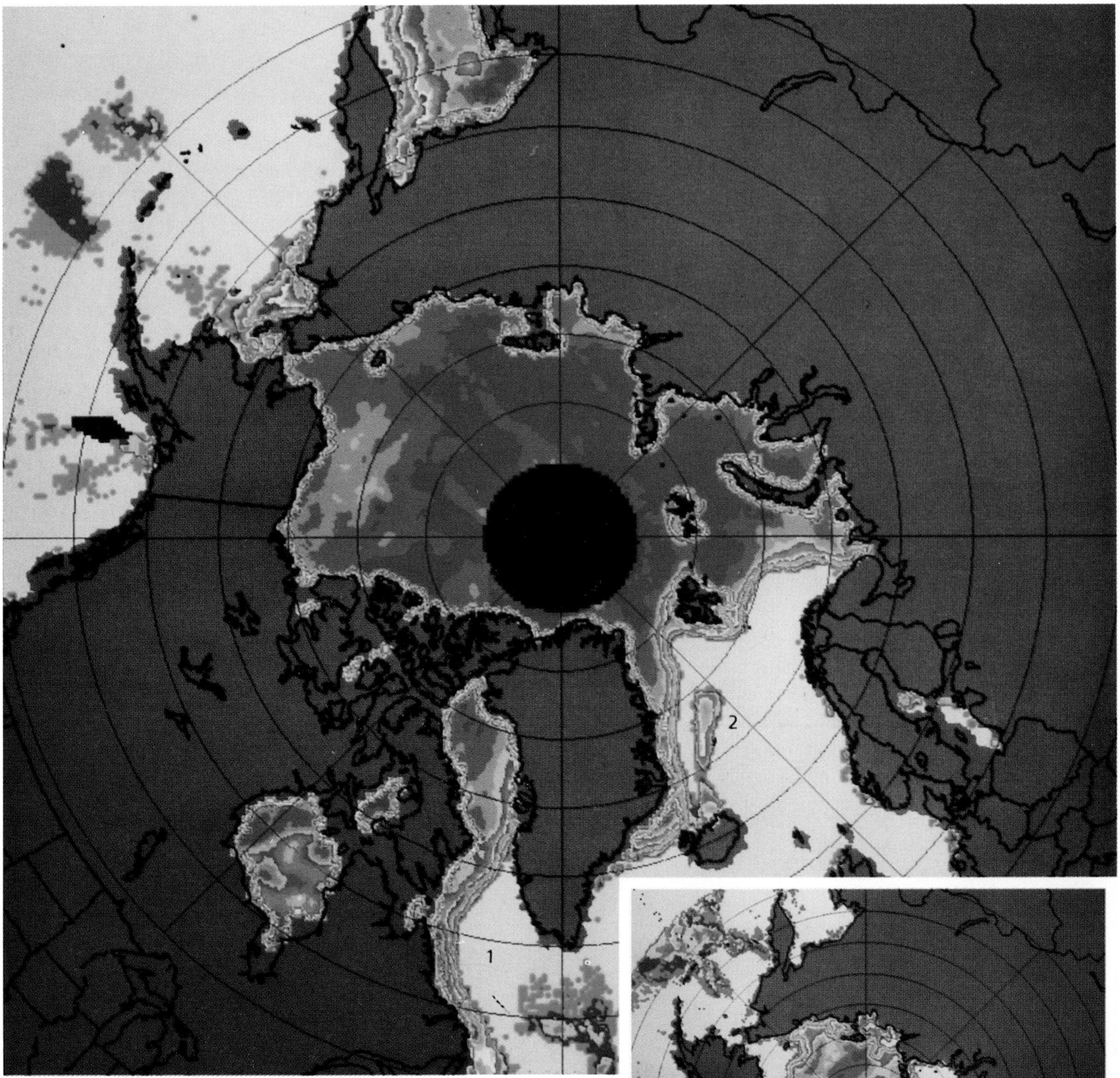

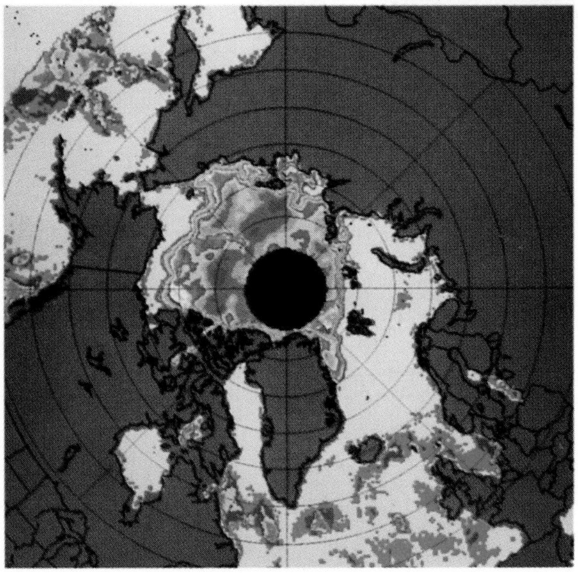

Figure 1.5 Satellite pictures showing typical seasonal changes in Arctic sea-ice cover.

In April, following the dark winter months, ice fills most of the Arctic Ocean, Hudson Bay and the Sea of Okhotsk. It also intrudes into the Bering Sea and along the coast of Greenland. The ice is restricted to a narrow tongue off Labrador (1) due to contact with warm Gulf Stream water from the south. East of Greenland a huge band of ice has separated from the main ice pack (2).

(Inset): In September, following warmer months of nearly continual daylight, the ice pack has melted and receded to the confines of the Arctic Ocean.

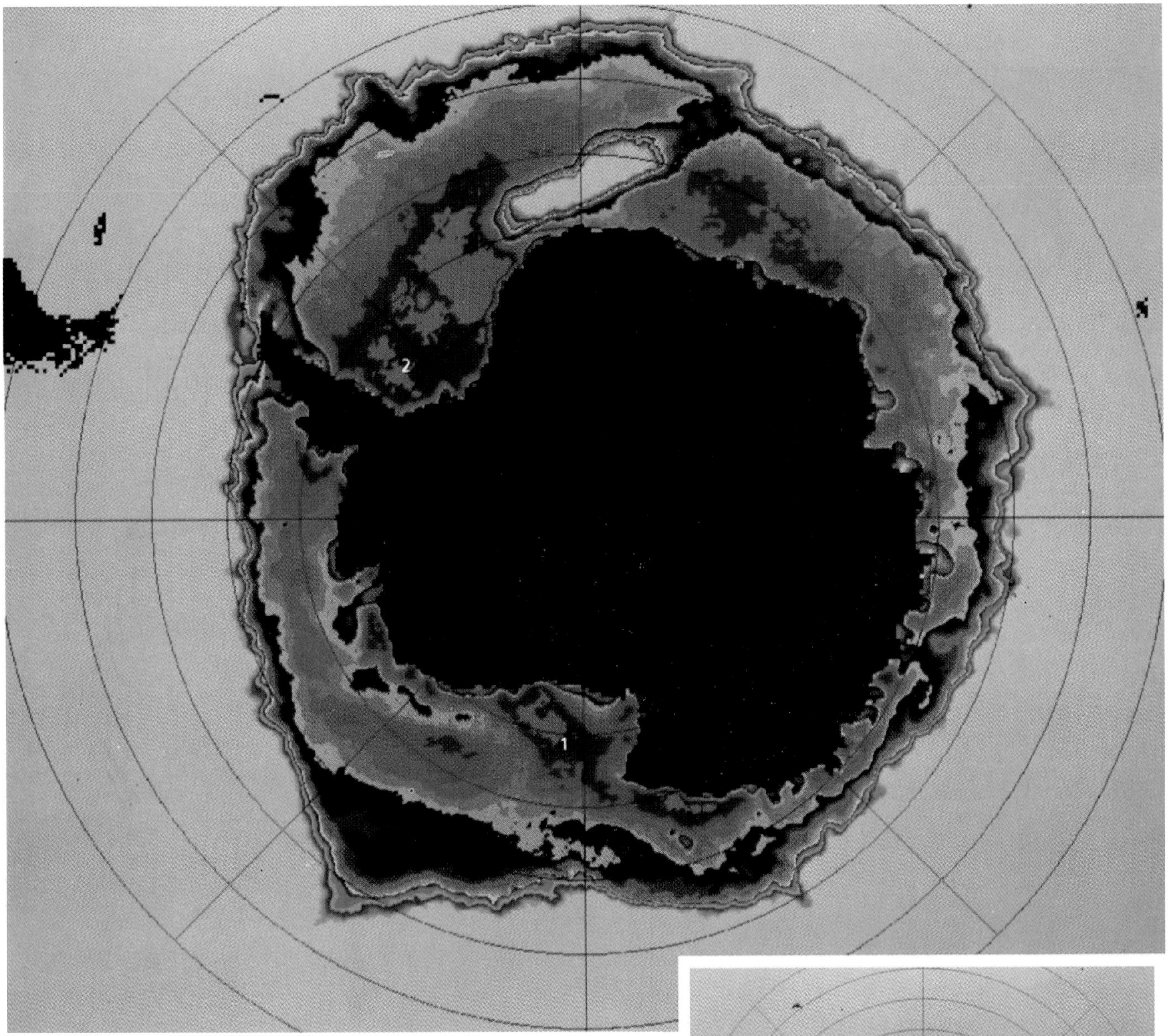

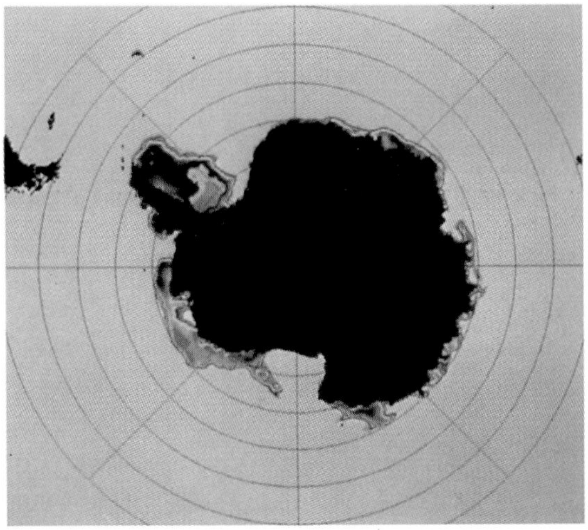

Figure 1.6 Satellite pictures showing typical seasonal changes in Antarctic sea-ice cover.

In August (the southern winter), the ice cover surrounds Antarctica, extending more than 1 000 km from the continent into the Ross (1) and Weddell (2) Seas. The large ice-free enclosure seen in the eastern part of the Weddell Sea is of particular interest. It does not form every year, but when it does, it is found in approximately the same position. Why it forms is not fully understood, but it must be due to warm water rising from below to ensure its survival through the winter.

(Inset): By February, sustained heating during continuous summer daylight has melted more than 80% of the winter ice cover, and ice extent is at its summer minimum.

Seasonal and inter-annual changes in ice cover at the poles can nowadays be monitored by satellite (Figures 1.5 and 1.6). A knowledge of the extent of ice cover and the way it is changing with time is crucial to understanding and predicting weather patterns. This is because ice and water reflect different amounts of solar radiation (see Table 2.1), and such data are important for assessing the surface radiation balance of the Earth as a whole.

Icebergs form in different ways in the two hemispheres. In the Arctic, they are mostly brought to the sea by valley glaciers from land masses such as Greenland and Spitzbergen, and they are irregular in shape. Thick ice-sheets fringe parts of the Arctic Ocean and large areas occasionally break off to form ice islands, upon which scientific bases have been established to make observation platforms that drift around the Arctic Ocean. In the Southern Ocean, tabular icebergs are more typically formed by 'calving' from the ice-shelves of the Ross and Weddell Seas (Figure 1.7).

(a)

(b)

Figure 1.7 (a) Seaward-facing cliffs of part of the Antarctic ice-shelf. The flow of ice from the land causes the ice-front to advance, but this is balanced by the 'calving' of (b) tabular icebergs from the cliffs.

Arctic icebergs contain more soil and debris eroded from the land by the glaciers and are usually denser and darker in colour than those of the Antarctic. They are also smaller, being rarely more than 1km long, though they can be as high as 60m above the sea-surface. Antarctic icebergs may have surface areas of many square kilometres, but they are rarely more than 35m high.

As icebergs melt, they dilute the surface seawater with freshwater, and the salt content of surface seawater in high latitudes is appreciably less than in ice-free latitudes: about 30–33$\,g\,kg^{-1}$, as against the average 35$\,g\,kg^{-1}$.

Everyone knows that it gets progressively colder as you travel from the Equator to the poles, and that when it is winter in the Northern Hemisphere it is summer in the Southern Hemisphere, and vice versa. However, not everyone would be able to explain just why these things happen. It is one of the topics considered in the next Chapter, which is concerned with the distribution of temperature in the oceans and some of the reasons for that distribution.

1.3 SUMMARY OF CHAPTER 1

1 The special properties of water—in particular, its anomalously high melting and boiling points, specific and latent heats, powerful solvent properties, and maximum density at 4°C—result from the polar structure of the water molecule. Dissolved salts increase the density of water and depress both the temperature of maximum density and the freezing point.

2 The oceans contain 97% of the water that circulates in the hydrological cycle. The residence time of water in the oceans is measured in thousands of years; in the atmosphere, it is measured in days.

3 Air is saturated with water vapour when evaporation is balanced by condensation. Clouds and fog are condensed water vapour. Fog may form when air is cooled to its condensation temperature, either by radiation from the land, or by advection of warm moist air over a cool land or water surface.

4 Sea-ice is less saline than the seawater from which it freezes, so its formation increases the salt content of the remaining seawater, thus further depressing its freezing point and increasing its density. Icebergs in the Northern Hemisphere are formed when valley glaciers on lands surrounding the Arctic Ocean reach the sea; those in the Southern Hemisphere break off from the thick ice-shelf that surrounds the Antarctic continent.

Now try the following questions to consolidate your understanding of this Chapter.

QUESTION 1.5 (a) In what ways are the thermal properties of water probably the single most important factor in preventing extremes of temperature from being reached at the Earth's surface?

(b) Most liquids reach a maximum density at freezing point, but pure water is an exception. At what temperature does pure water reach a maximum density? Does this temperature apply to seawater?

QUESTION 1.6 What is the approximate average residence time of water on land, and why is this average value likely to conceal considerable variations?

QUESTION 1.7 (a) What is the main difference in origin between the ice-sheets that cover the Arctic and Antarctic regions?

(b) A sample of water contains 20g kg^{-1} dissolved salts. At what temperature will it (i) attain its maximum density, (ii) freeze?

(c) Ice melts and mixes with seawater of salinity 35g kg^{-1}. Will this have the effect of raising or lowering the freezing point of the seawater? Would this in turn tend to facilitate the formation of more sea-ice when temperatures fell once more?

CHAPTER 2 TEMPERATURE IN THE OCEANS

Two of the most important physical properties of seawater are temperature and salinity (salt content), for together they control its density, which is the major factor governing the vertical movement of ocean waters.

In the oceans, the density of seawater normally increases with depth. If the density of surface water increases, it becomes gravitationally unstable and sinks. In polar regions, the density of surface waters can be increased in two ways: first, by direct cooling, either where ice is in contact with the water or where cold winds blow off the ice; secondly, by the formation of sea-ice, which extracts water and leaves behind seawater of higher salinity, and increased density. The cold dense currents of the deep circulation (see Chapter 4) originate by sinking of dense water in polar regions. In lower latitudes, dense saline water is produced by excess evaporation, which may be aided by strong winds such as those that occur during the winter in parts of the Mediterranean.

2.1 SOLAR RADIATION

The Sun's radiation is dominated by ultraviolet, visible and near infrared wavelengths (see Section 2.3). On average, only about 70% of the solar radiation that reaches the Earth penetrates the atmosphere. About 30% (on average) is *reflected* back into space from clouds and dust particles. Of the remaining 70%, on average:
about 17% is *absorbed* in the atmosphere;
about 23% reaches the surface as *diffuse* daylight;
about 30% reaches the surface as *direct* sunlight.

Much of the ultraviolet radiation is absorbed in the ozone layer. (A cloudless sky appears blue because of scattering of shorter wavelengths by molecules of air.)

The radiation that actually reaches the Earth's surface—the **insolation**—is not all absorbed. The percentage of the insolation reflected by a surface is called the **albedo** of that surface. Some typical terrestrial albedos are given in Table 2.1, from which it is evident that polar ice-caps absorb only a relatively small proportion of the insolation. Waves and ripples significantly increase the albedo of water, but it is still generally less than that of most surfaces on land. The time of day will also affect the albedo (especially of water, ice or snow), because the shallower the angle of incidence of the solar radiation the greater the amount reflected.

Table 2.1 Some typical albedos.

Surface	Albedo (%)
Snow	up to 90
Desert sands	35
Vegetation	10–25
Bare soil or rock	10–20
Built-up areas	12–18
Calm water	2

QUESTION 2.1 At the same latitude, season and time of day, would you expect the atmosphere above a sandy beach to be warmed more or less than that above a pine wood?

Some of the radiation reflected back from the Earth's surface is absorbed in the atmosphere and warms it further. The surface is warmed by the radiation it has absorbed, and in turn radiates back infrared and longer (microwave) wavelengths. Atmospheric water vapour and CO_2 strongly

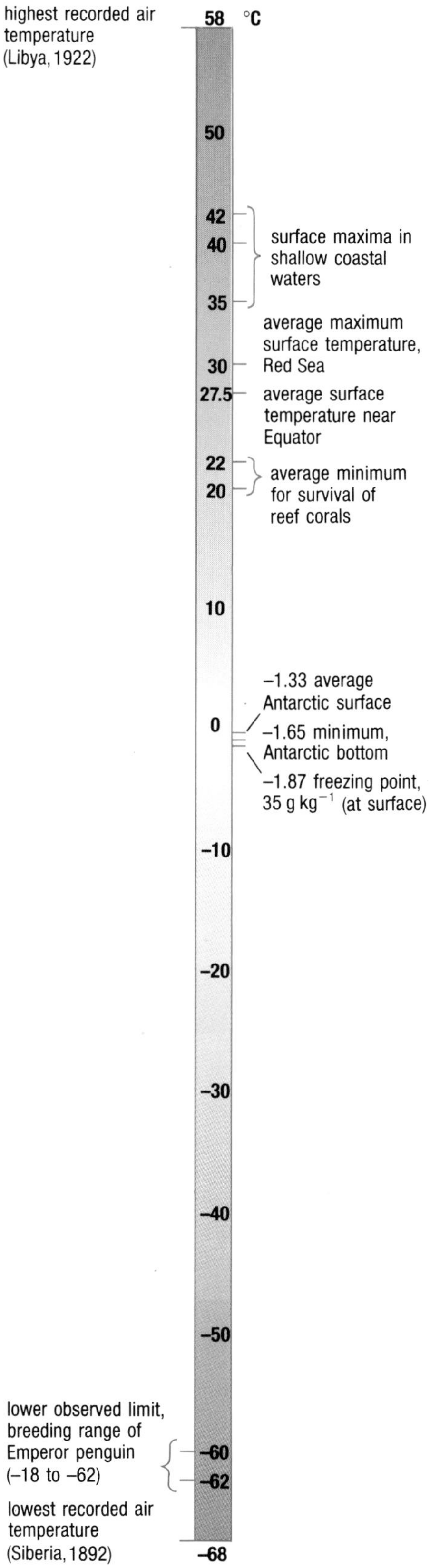

Figure 2.1 Temperature ranges in the sea (right) and on land (left).

absorb infrared wavelengths, and so the atmosphere acts very much as a blanket to keep in the heat. This has been called the **greenhouse effect** (probably incorrectly, because greenhouses mainly trap heat by preventing heated air from escaping by convection), and it has been known for some time that the increase in atmospheric CO_2 from the burning of fossil fuels such as coal and petroleum is contributing to a general long-term warming of the atmosphere. Carbon dioxide and water vapour are not the only 'greenhouse gases'. Other contributors to the effect are methane and nitrous oxide (see Chapter 6), and the artificially produced chlorofluorocarbons (e.g. CCl_3F), which are used as refrigerants and commercial aerosol propellants, and are now causing concern because they are implicated in the depletion of ozone in the upper atmosphere. The stratospheric 'ozone hole' observed over Antarctica became a well-publicized environmental issue in the mid-1980s, because of the fear that more ultraviolet radiation will reach the Earth's surface, with harmful consequences to life.

Diurnal (daily) variations of temperature on land are sometimes measurable in tens of degrees, but in the oceans they amount to no more than a few degrees, except in very shallow water.

QUESTION 2.2 With the help of Table 1.1, can you suggest three main reasons for this?

The answers to Question 2.2 account also for the contrasts displayed in Figure 2.1: the range of temperature in the oceans is about 40°C (or about 30°C if we exclude shallow and restricted seas); whereas the temperature range encountered on land is about three times greater. The temperature-buffering effect of the oceans (Question 1.5(a)) depends on the continuous exchange of heat and water between ocean and atmosphere, mainly by means of the hydrological cycle (Figure 1.3).

2.2 DISTRIBUTION OF SURFACE TEMPERATURES

The intensity of insolation depends primarily on the angle at which the Sun's rays strike the surface (Figure 2.2(a)), and the distribution of temperature over the surface of the Earth varies with latitude and season, because of the tilt of the Earth's axis with respect to its orbit round the Sun. Figure 2.2(b) shows that along the Equator, maximum insolation occurs at the March and September **equinoxes**, when the noonday Sun is overhead. Insolation remains high in equatorial regions during the rest of the year. The noonday Sun is overhead along the Tropics of Cancer and Capricorn at the June and December **solstices** respectively, so temperate latitudes receive maximum and minimum insolation during their respective summer and winter seasons. There is insolation for only about half the year at the poles which are wholly illuminated in summer and wholly dark in winter.

Until the advent of satellite technology, it was impossible to monitor seasonal changes of sea-surface temperature over wide areas. Satellite-mounted infrared sensors now make it possible to measure the change of sea-surface temperature on a global scale, both seasonally and from year to year (Figure 2.3). The sensitivity and precision of the sensors is of the order of ±0.1°C, but their absolute accuracy is less satisfactory, because of the errors induced by factors such as the state of the sea-surface (smooth or rough) and the amount of water vapour in the atmosphere (water vapour absorbs infrared radiation). Nonetheless, information such as that in Figure 2.3 can be obtained on a more or less continuous basis, and for many oceanographic purposes it is the *variation* in sea-surface temperature that is important, rather than the absolute values. It is essential to bear in mind that this is information about the sea-surface *only*. Satellite-based instruments cannot discover anything about the depth-related temperature structure of the oceans (see also Chapter 5).

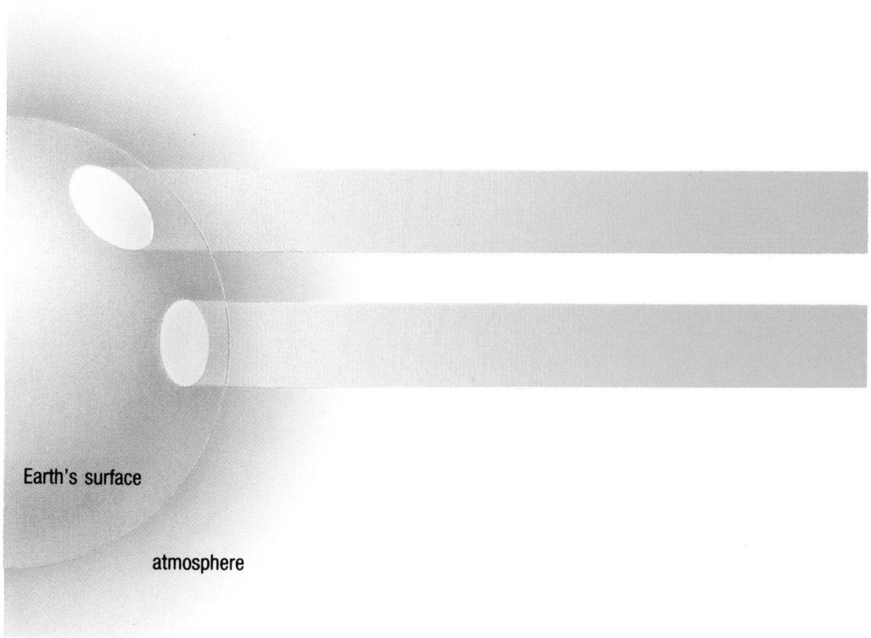

(a)

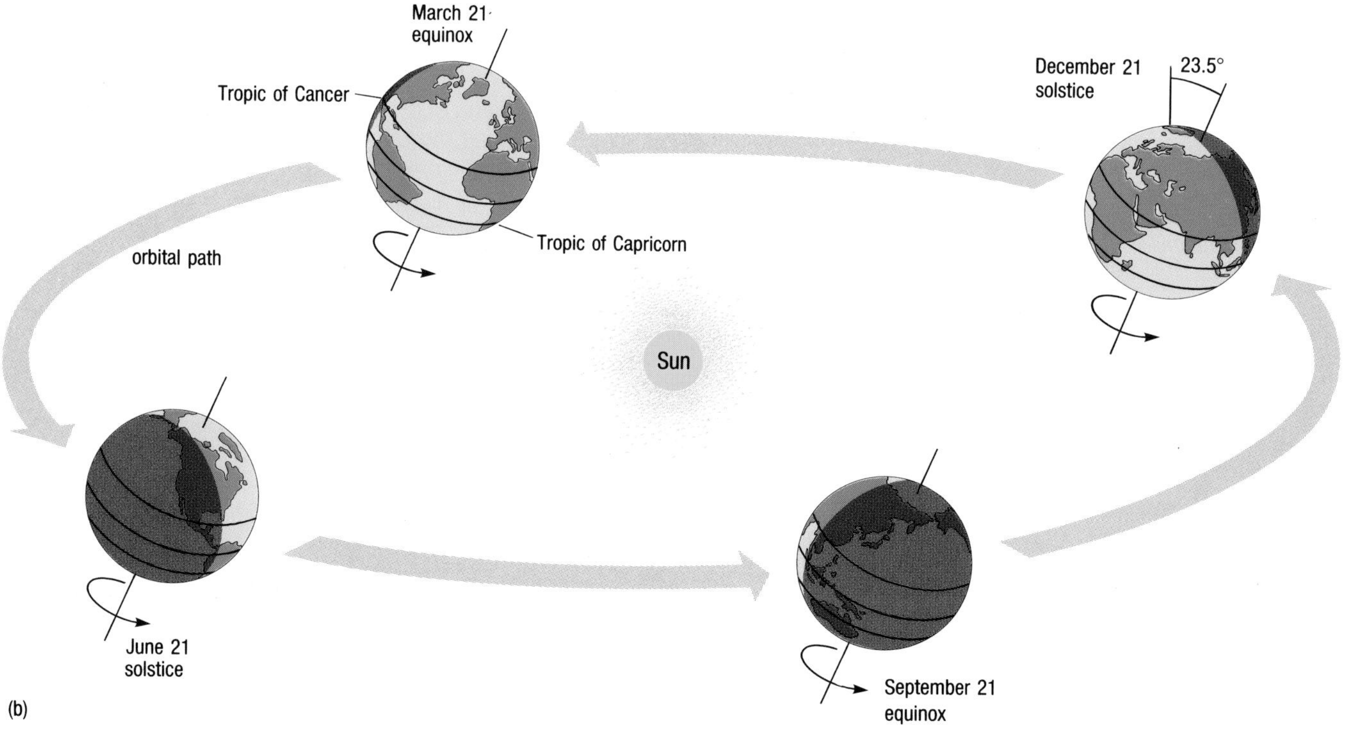

(b)

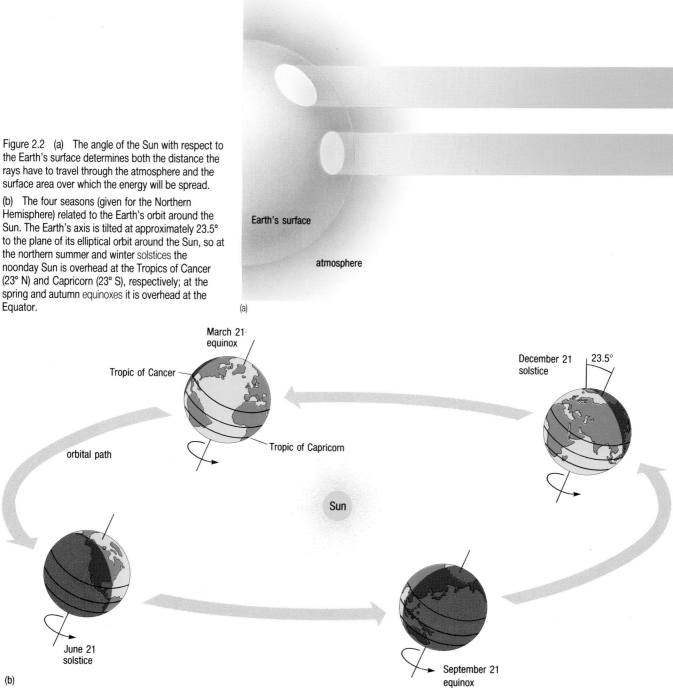

Figure 2.2 (a) The angle of the Sun with respect to the Earth's surface determines both the distance the rays have to travel through the atmosphere and the surface area over which the energy will be spread.

(b) The four seasons (given for the Northern Hemisphere) related to the Earth's orbit around the Sun. The Earth's axis is tilted at approximately 23.5° to the plane of its elliptical orbit around the Sun, so at the northern summer and winter solstices the noonday Sun is overhead at the Tropics of Cancer (23° N) and Capricorn (23° S), respectively; at the spring and autumn equinoxes it is overhead at the Equator.

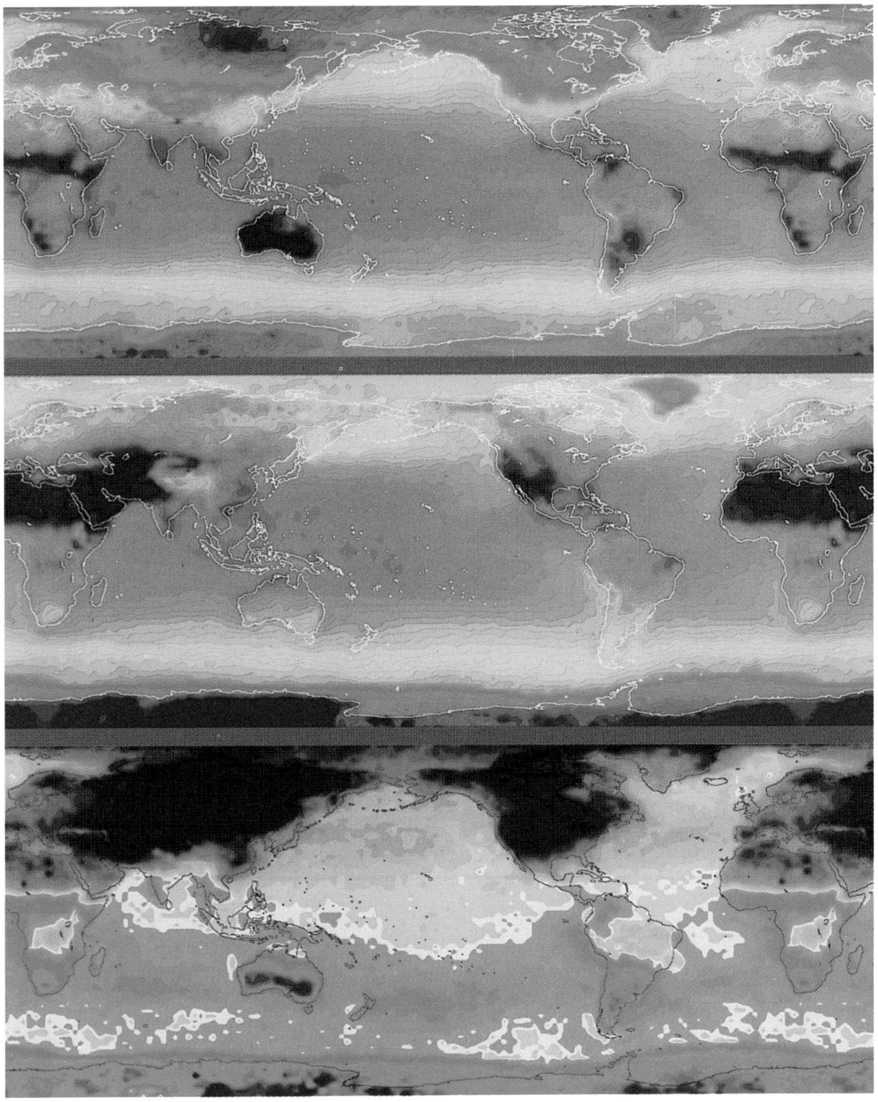

Figure 2.3 Sea-surface temperature measurements from satellites. Temperatures are on the absolute Kelvin scale (273 Kelvin = 0° Celsius). For the upper two pictures, temperatures below freezing (273 K) are green and blue. Higher temperatures are red and brown.

(*Top*): In January, the Northern Hemisphere experiences extreme cold. In Siberia and over most of Canada, temperatures approach −30° C, while in Eastern Europe and the northern USA temperatures are below 0° C. In the Southern Hemisphere it is summer, with mid-latitude temperatures ranging from 20° C to 30° C. On the eastern and western sides of the open oceans, the contours of equal temperature show deviations from their latitudinal (or zonal) patterns. Generally, in the sub-tropics of both hemispheres (10° to 30°), the western sides of the oceans are warmer than their eastern counterparts, primarily due to ocean currents. The Gulf Stream can be seen moving along the North American continent, then turning north-eastwards and transporting warm waters across the Atlantic to moderate the climate of north-west Europe.

(*Middle*): By July, areas of the Northern Hemisphere have warmed to 10° C to 20° C. Equatorial Africa and India are the hottest. In the Arctic, Greenland remains frozen, while Hudson Bay has thawed. In the Southern Hemisphere, Antarctica is much colder than the Arctic and ice has formed in the Weddell Sea.

(*Bottom*): This shows temperature *differences* between January and July, and emphasizes the point made in Figure 2.1 and Question 2.2. The greatest warming and cooling has occurred over land (dark blue, brown). Marked seasonal changes of up to 30° C are seen on land in both hemispheres. In contrast the changes in ocean temperature rarely exceed 8° C to 10° C. The greatest deviations are in mid-latitudes, while the near-equatorial regions are quite stable. In the Northern Hemisphere, mid-latitude changes in ocean temperature are strongly influenced by the position of continents. The continents divert the ocean currents and affect wind patterns. In the Southern Hemisphere, which has only half the land area of the Northern Hemisphere, changes are primarily due to seasonal variations of incoming solar radiation.

2.2.1 THE TRANSFER OF HEAT AND WATER ACROSS THE AIR–SEA INTERFACE

The surface temperature of the sea depends on the insolation it receives, and on the amount of heat it radiates back into the atmosphere: the warmer the surface, the more heat it radiates. Heat is also transferred across the surface of the sea by conduction and by the effects of evaporation.

Conduction

If the sea-surface is warmer than the air directly above it, heat can be transferred from the sea to the air. On average, the sea-surface is warmer than the overlying air masses, so there is a net loss of heat from the sea by conduction. This loss is relatively unimportant in the total heat budget of the oceans, and it would be negligible were it not for the effect of convective mixing by wind, which removes the warmed air from just above the sea-surface.

Evaporation

Evaporation (the transfer of water to the atmosphere) is the main mechanism by which the sea loses heat—about an order of magnitude more than is lost by conduction plus convective mixing.

The governing equation is:

(rate of loss of heat) = (latent heat of evaporation) × (rate of evaporation).

QUESTION 2.3 (a) Use Figure 1.3 to calculate an approximate value for the heat lost from the oceans by evaporation each day, using the value for latent heat of evaporation given in Table 1.1.

(b) Given that the Earth's surface receives about 9×10^{21} J from the Sun each day (70% of the incoming solar radiation, Section 2.1), would you say that evaporation from the oceans is a significant component in the Earth's heat budget?

(c) Under what conditions would you expect ocean water to *gain* heat by condensation?

Evaporation, condensation and precipitation are not the only mechanisms for transferring water across the interface of air and sea. As with all liquid bodies, the outer surface of the ocean is defined by intermolecular forces that cause a **surface tension**. The surface tension of seawater is less than that of freshwater, so seawater more readily breaks into froth or foam when disturbed. High winds cause foaming and streaking of the surface layers, as well as entrapment of air bubbles.

Figure 2.4(a) shows what happens when air is injected into subsurface water under rough conditions, with breaking waves and white caps. Bubbles of trapped air rise to the surface and break, injecting droplets of various sizes into the atmosphere, along with any dissolved salts, gases and particulate matter that the water may contain. A large proportion of these constituents is soon returned to the surface by precipitation, as shown by the decrease in chloride content of rainwater with increasing distance inland from coasts (Figure 2.4(b)). The smallest droplets injected

Figure 2.4 (a) Diagrammatic representation of the successive stages of bubble collapse, for a 'typical' 1 mm diameter bubble. (μm = micrometre (micron) = 10^{-6} m and ng = nanogram = 10^{-9} g.)

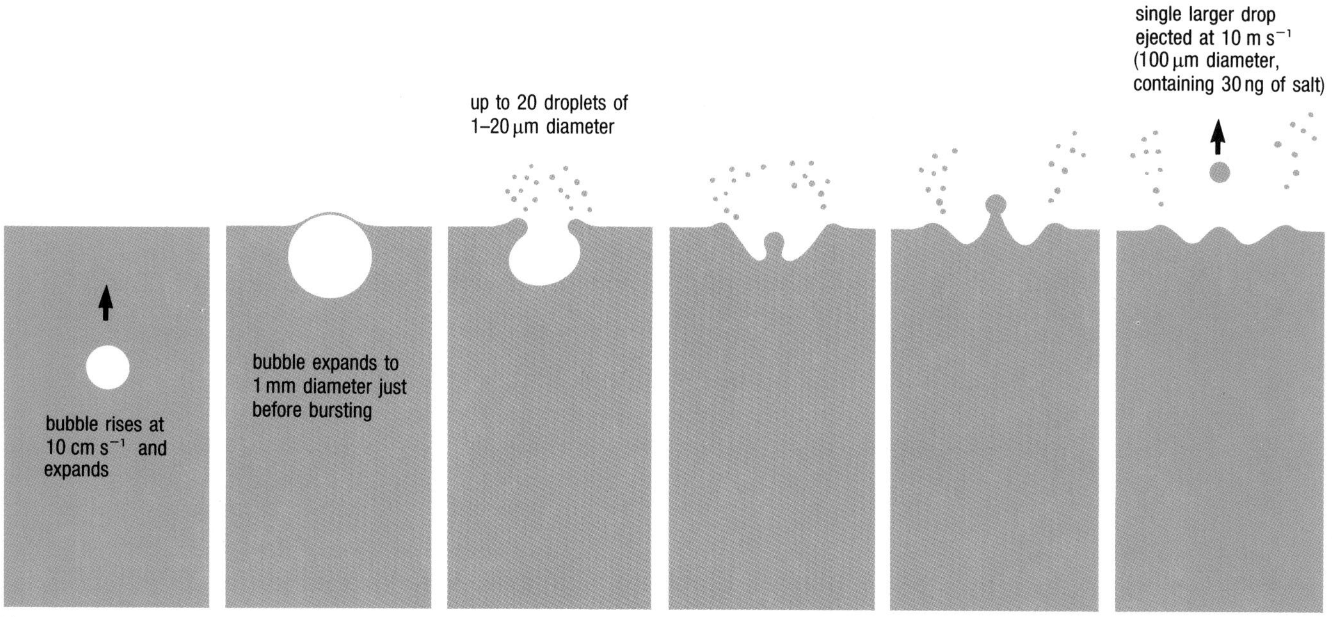

(a)

Figure 2.4 (b) Decrease in chloride content of rainwater with increasing distance inland from the coast.

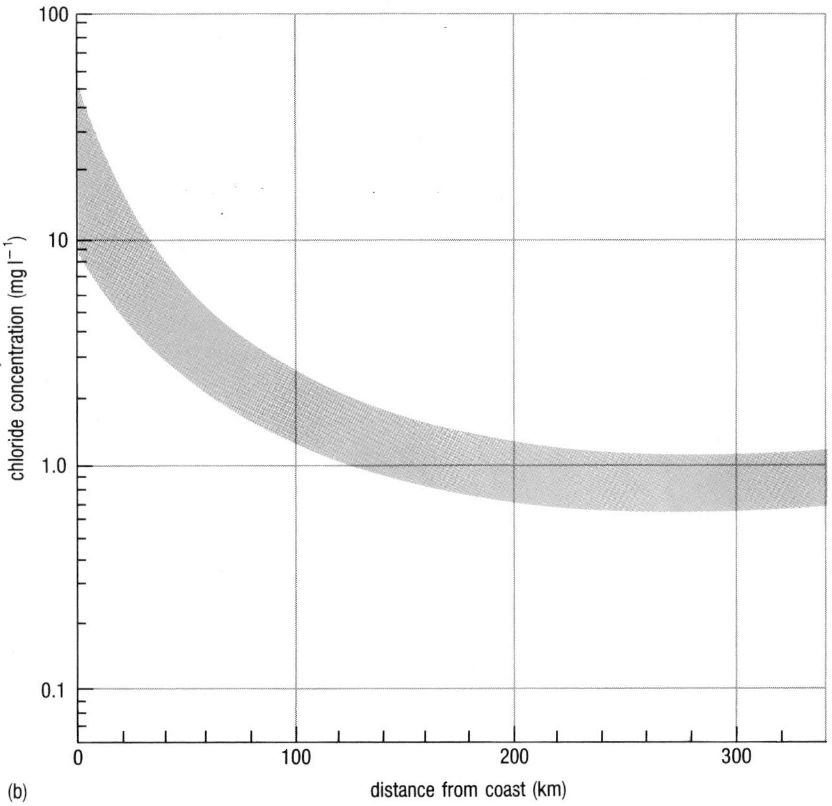

into the atmosphere are called **aerosols**, and they remove water, dissolved salts and organic matter from the surface of the oceans. Aerosols can be carried high above the Earth and dispersed throughout the atmosphere. When the water evaporates, the minute precipitated particles of salt and other substances act as nuclei for cloud and rain formation.

2.3 DISTRIBUTION OF TEMPERATURE WITH DEPTH

Measurement of temperature at the surface of the ocean, let alone below it, was not possible until the thermometer was invented in the early 17th century. The earliest temperature measurements were made on water samples collected in iron or canvas buckets from surface waters. It was realized that temperature decreased with depth, but accurate measurement of subsurface temperatures became possible only when thermometers, protected against the water pressure and capable of recording *in situ* temperatures, were invented in the mid-nineteenth century, shortly before the voyage of HMS *Challenger*. Such problems no longer arise, because temperature in the oceans is nowadays measured with thermistors, and continuous temperature profiling—both vertical and lateral—is now a routine oceanographic procedure.

Figure 2.5 shows that most solar energy is absorbed within a few metres of the ocean surface, which directly heats the surface waters and provides the energy for the formation of organic molecules by photosynthesis.

Why is the colour of the light below the surface of the sea predominantly blue–green? Which wavelengths are the first to be absorbed? What proportion of the total incident energy reaches a depth of 100m?

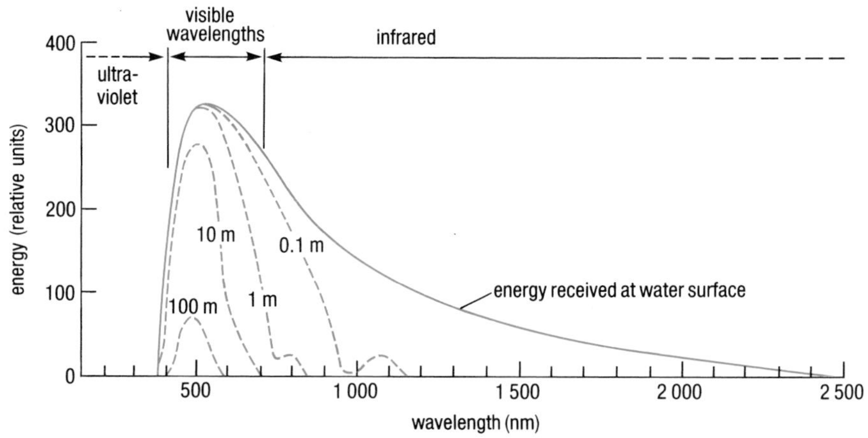

Figure 2.5 A simplified energy-wavelength spectrum of solar radiation at the surface of the ocean and at various depths. (nm = nanometre = 10^{-9}m.)

Shorter wavelengths, i.e. those near the blue end of the visible spectrum, penetrate deeper than longer wavelengths. Infrared radiation is the first to be absorbed, followed by red, and so on. The total energy received at a given depth is represented by the area beneath the relevant curve on Figure 2.5. The different areas beneath the curves for 100m and the water surface suggest that only about one-fiftieth of the incident energy penetrates to 100m. All of the infrared radiation is absorbed within about a metre of the surface, and nearly half of the total incident solar energy is absorbed within 10cm of the surface. Penetration will also depend on the clarity or transparency of the water, which in turn depends on the amount of suspended matter in it (see also Chapter 5).

If the thermal energy from solar radiation is largely absorbed by the surface layers, how can it be carried deeper?

Conduction by itself is extremely slow, so only a small proportion of heat is transferred downwards by this process. The main mechanism is turbulent mixing by winds and waves, which establishes a **mixed surface layer** that can be as thick as 200–300m at mid-latitudes in the open oceans, whereas it might be as little as 10m thick in sheltered coastal waters in summer.

Between about 200–300m and 1000m depth, the temperature declines rapidly throughout much of the ocean. This region of steep temperature gradient is known as the **permanent thermocline**, beneath which, from about 1000m to the ocean floor, there is virtually no seasonal variation and (except in polar regions) temperature decreases gradually to between about 0°C and 3°C (Figure 2.6(a)). This narrow range is maintained throughout the deep oceans, both geographically and seasonally, because it is determined by the temperature of cold, dense water that sinks from the polar regions and flows towards the Equator along the ocean floor.

QUESTION 2.4 Figure 2.6(a) is a vertical section illustrating the range of temperatures encountered in the oceans, and Figure 2.6(b) shows vertical temperature–depth profiles along lines A and B in Figure 2.6(a).

(a) Match the profiles I and II in Figure 2.6(b) with vertical lines A and B in Figure 2.6(a).

(b) What can you say about the vertical distribution of temperature at high latitudes (above about 60°N and 60°S)?

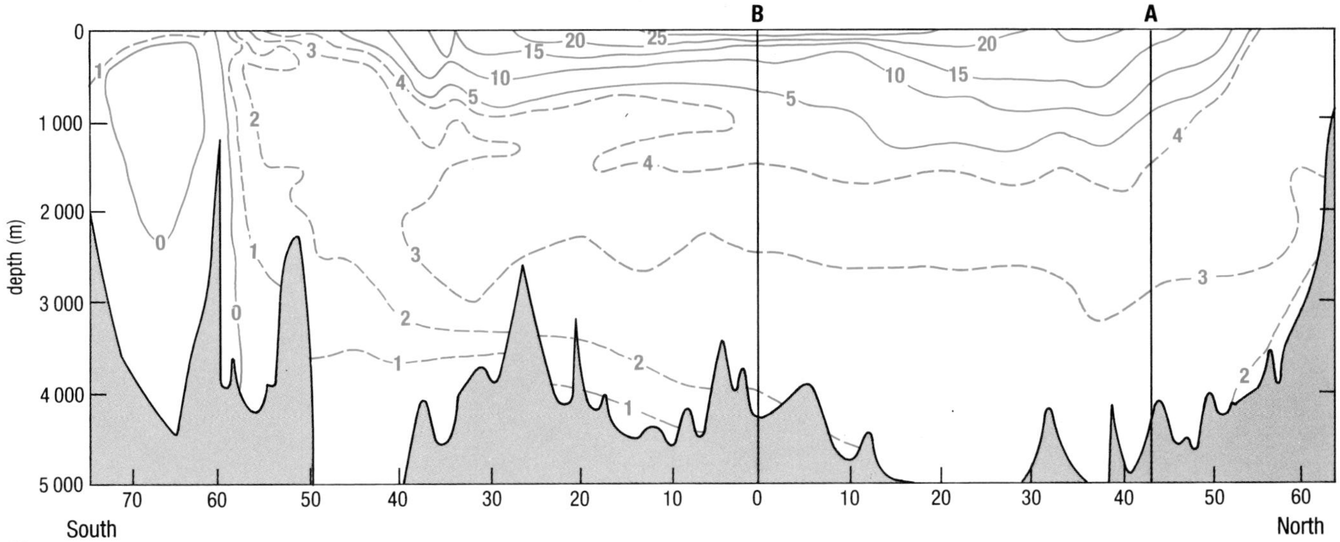

(a)

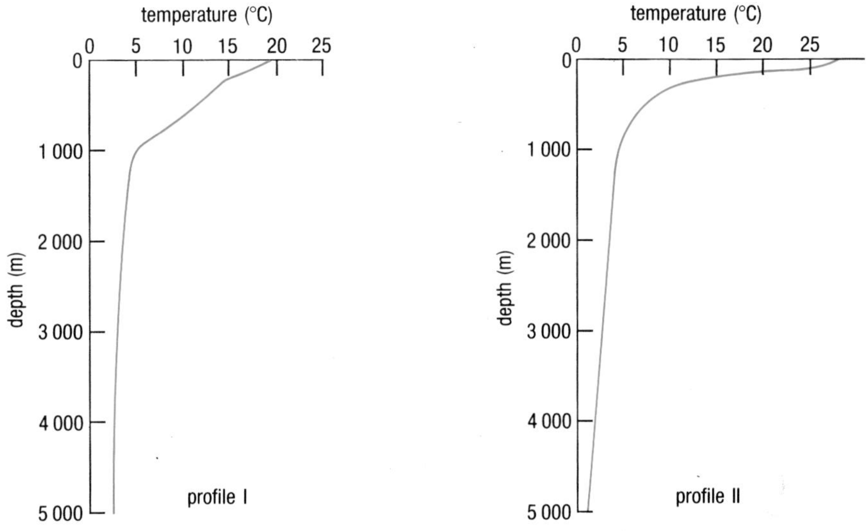

(b)

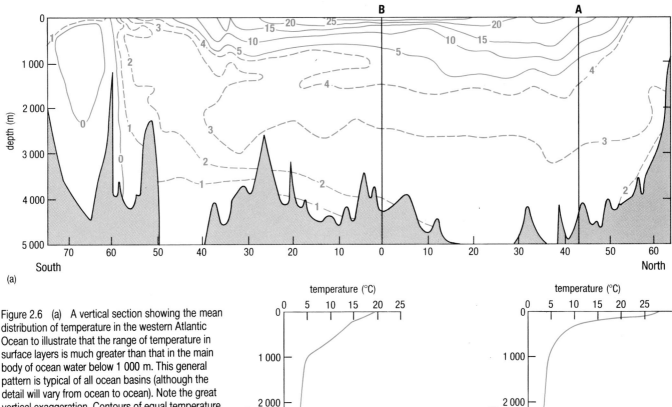

Figure 2.6 (a) A vertical section showing the mean distribution of temperature in the western Atlantic Ocean to illustrate that the range of temperature in surface layers is much greater than that in the main body of ocean water below 1 000 m. This general pattern is typical of all ocean basins (although the detail will vary from ocean to ocean). Note the great vertical exaggeration. Contours of equal temperature are called **isotherms**. Broken line isotherms 1 °C interval; solid lines 5 °C interval. The vertical lines A and B relate to Figure 2.6 (b) and are for use with Question 2.4.

(b) Temperature profiles along A and B in (a), for use with Question 2.4.

The temperature and depth of the mixed surface layer show seasonal variations in mid-latitudes. During the winter, when surface temperatures are low and conditions at the surface are rough, the mixed surface layer may extend to the permanent thermocline; i.e. the temperature profile can be effectively vertical through the top 200–300m or more. In summer, as surface temperatures rise and conditions at the surface are less rough, a **seasonal thermocline** often develops above the permanent thermocline, as shown in the generalized profile of Figure 2.7(a).

Seasonal thermoclines start to form in spring and reach their maximum development (i.e. with steepest temperature gradients) in the summer. They are usually at depths of a few tens of metres, with a mixed layer above (Figure 2.7(a)). Winter cooling and strong winds progressively increase the depth of seasonal thermoclines and reduce the temperature gradient along them, so that the mixed surface layer reaches its full thickness of 200–300m (see Figure 2.7(d)). In low latitudes, there is no

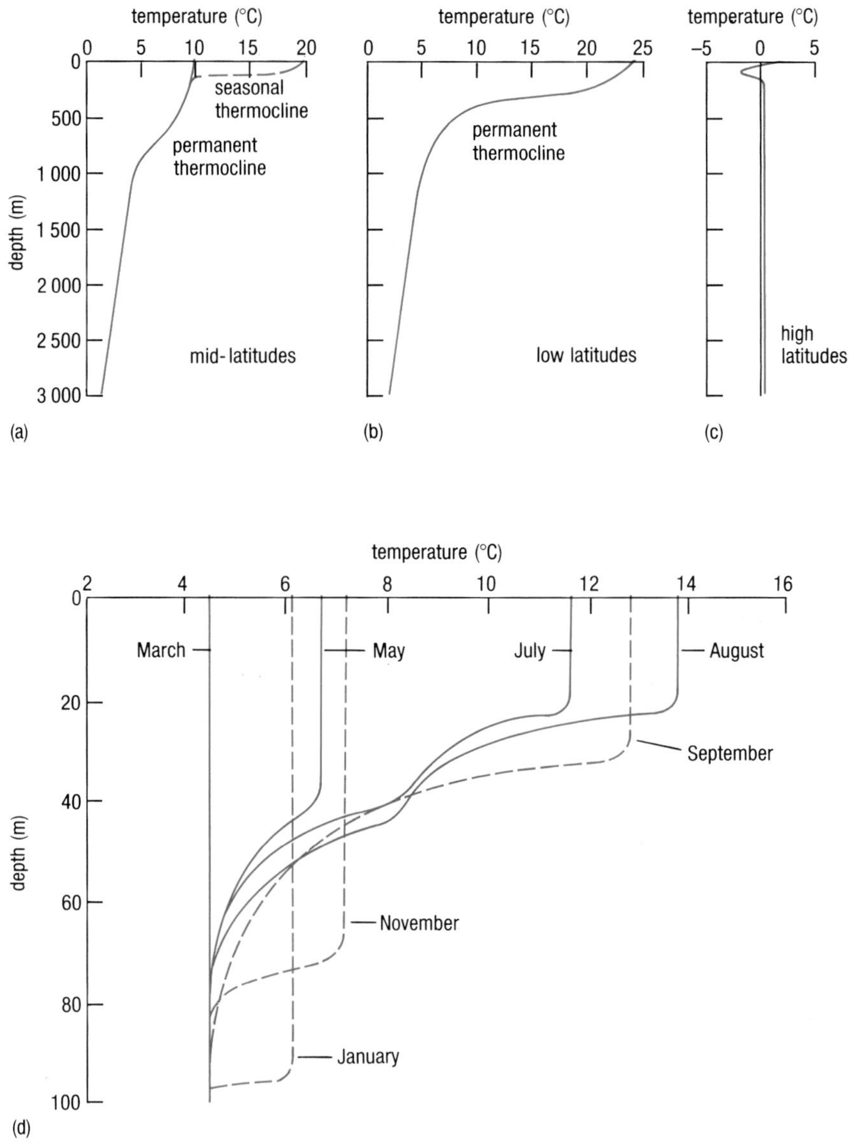

Figure 2.7 (a)–(c) Typical mean temperature profiles for different latitude belts in the open oceans. Note that the otherwise vertical profile (c) for high latitudes shows a layer of colder water at 50–100 m, depth. (d) A succession of temperature profiles to show the growth (solid lines) and decay (broken lines) of a seasonal thermocline in the Northern Hemisphere. Note the very different scales compared to (a)–(c).

winter cooling, so the 'seasonal thermocline' becomes 'permanent' and merges with the permanent thermocline at depths of 100–150m (Figure 2.7(b)). At high latitudes greater than about 60° there is no permanent thermocline (Figure 2.7(c)). Nonetheless, seasonal thermoclines can still develop in summer at these higher latitudes, above the weak or non-existent permanent thermocline.

Figure 2.8 gives an idea of the pattern of temperature change with depth and season; the annual range of nearly 10°C at the surface declines to only about 3–4°C at 100m depth.

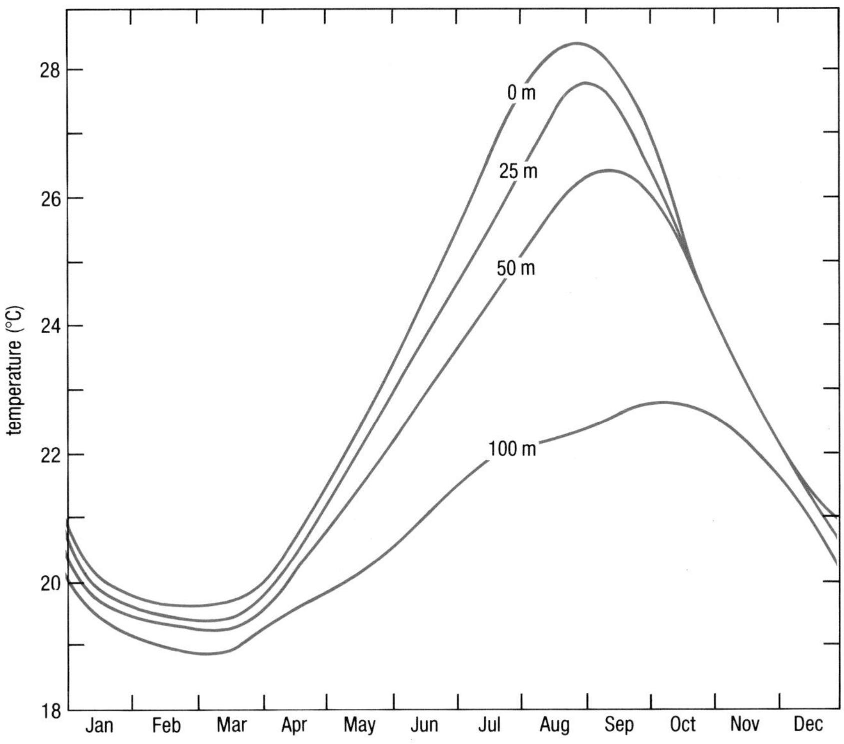

Figure 2.8 Annual variations of water temperature at different depths in the ocean off the south coast of Japan.

QUESTION 2.5 Given that changes of temperature resulting directly from seasonal variations of incident radiation can no longer be detected below about 200m, approximately where would you place the curve for 200m on Figure 2.8, and what form would you expect it to have?

Diurnal thermoclines can form anywhere, provided there is enough heating during the day, though they occur only at depths down to about 10–15m, and temperature differences across them do not normally exceed 1–2°C.

In summary, and ignoring seasonal and diurnal variations, the permanent thermocline allows the oceans as a whole to be divided into three principal layers, shown schematically in Figure 2.9. The depths of both mixed surface layer and permanent thermocline are less at low latitudes than at mid-latitudes because winds are generally weaker and seasonal temperature contrasts are less at low latitudes.

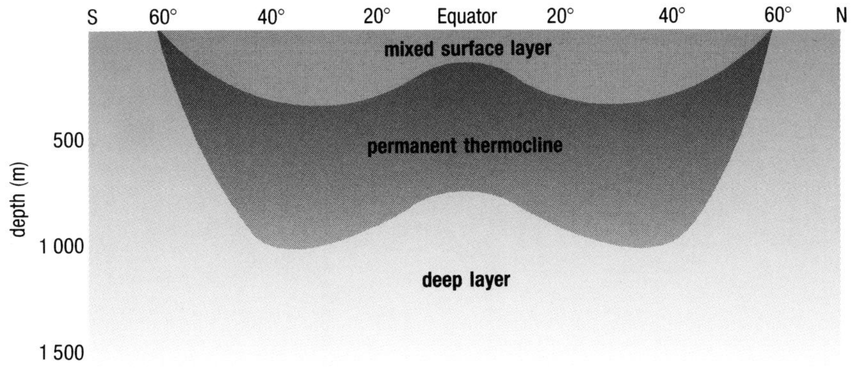

Figure 2.9 Generalized and schematic cross-section, showing the main thermal layers of the oceans.

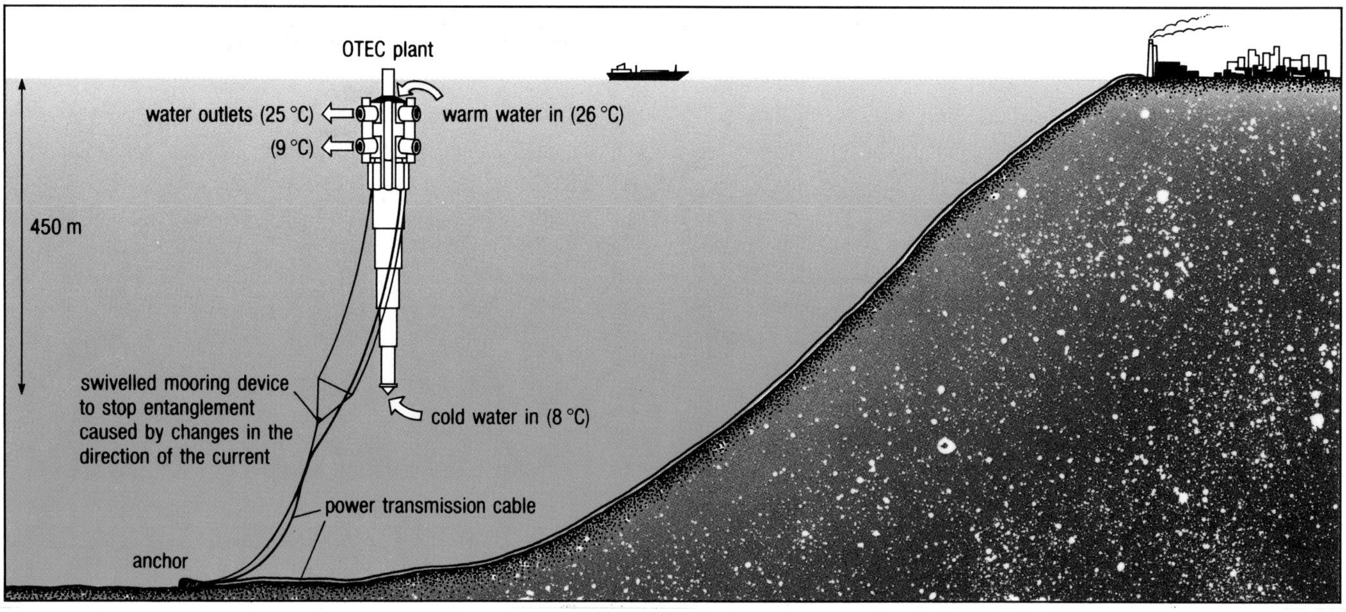

Figure 2.10 A general view of a proposed OTEC plant which would use the difference in temperature between the surface of the ocean and deep waters to obtain electrical energy. As much as 160 000 kW might be generated by such a plant.

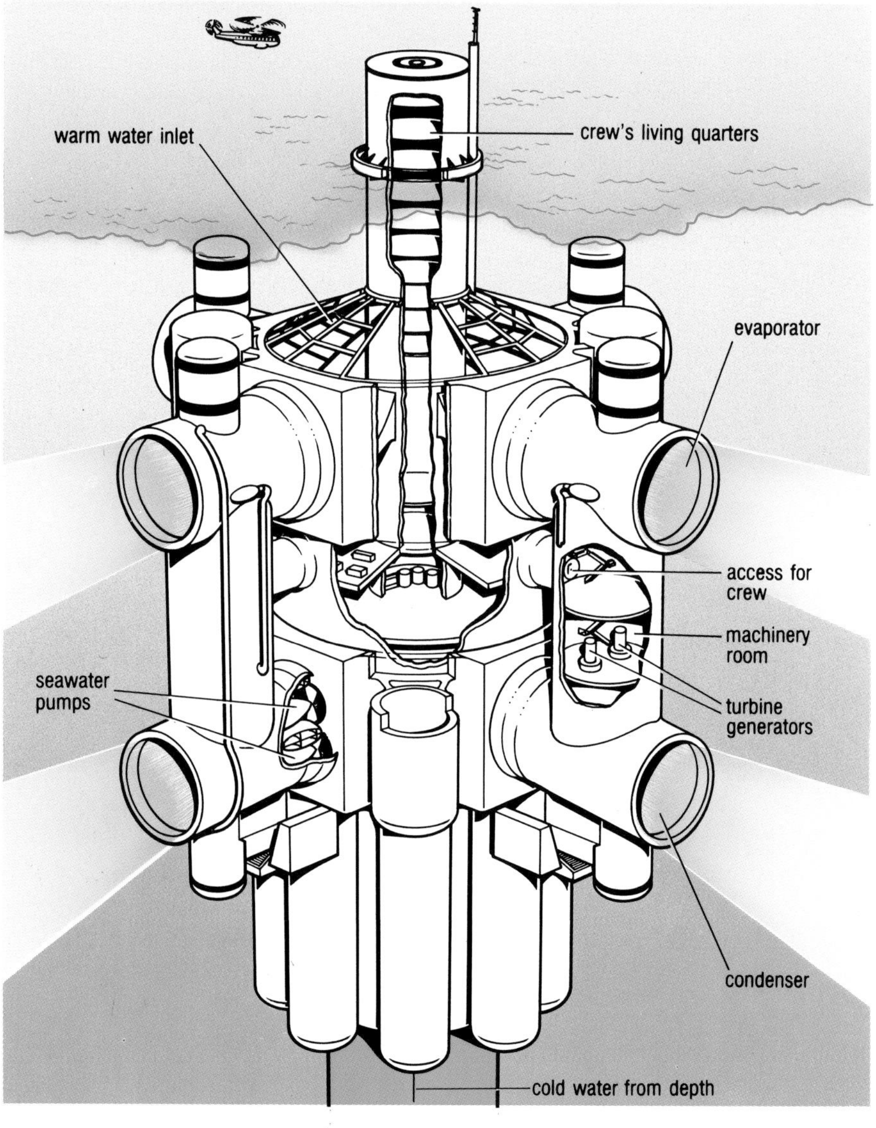

Figure 2.11 Details of the 'operating head' of an OTEC plant. Warm surface water at 26 °C would be passed through the evaporators and used to vaporize propane or ammonia. This vapour would be used to drive turbines and thus generate electricity before being passed through condensers and cooled by cold deep water.

2.4 ENERGY FROM THE THERMOCLINE—A BRIEF DIGRESSION

The permanent thermocline is found nearly everywhere in the oceans (Figure 2.9) and in low latitudes the temperature difference across it is of the order of 20°C, and sometimes more (Figures 2.6 and 2.7). The problems of tapping energy from this temperature gradient in ocean waters are mainly those of scale. The principle of Ocean Thermal Energy Conversion (**OTEC**) is exactly the same as that used in refrigerators, air conditioners and heat pumps. Warm surface water at about 25°C is pumped into one bank of heat exchangers to vaporize ammonia or freon which expands to drive turbines. At the same time, cold water at about 4°C from deeper levels in the oceans (below about 1000m), is pumped up to cool and condense the vapour in another bank of heat exchangers, continuing the cycle.

By the late 1980s the Japanese had advanced the furthest with this technology and had built some small plants generating between 50 and 100kW. At this scale, the most likely beneficiaries of the technology seem to be the small islands of the South Pacific—the island of Nauru on the Equator already has experience of operating a 100kW experimental plant. For larger OTEC power stations, huge installations would be required, moored in water depths of over 1000m and comparable in size to oil production platforms on continental shelves (Figures 2.10 and 2.11). Such enterprises are unlikely to be economic while supplies of relatively cheap fossil fuels last, although they are quite feasible technologically.

2.5 TEMPERATURE DISTRIBUTION AND WATER MOVEMENT

Sections and profiles such as those of Figures 2.6 and 2.9 represent temperatures averaged over periods of months or years. We know that large seasonal changes of temperature occur in the surface layers (e.g. Figure 2.8), and there can be small fluctuations with time, even in the deep oceans.

It is more important, however, not to gain the impression from such time-averaged temperature sections and profiles that the waters of the oceans are static. Far from it. It is essential always to bear in mind that while the locations of mean isotherms along such sections do not change significantly even on time-scales of decades, the structure is maintained dynamically. Any given parcel of water can travel a distance equivalent to a global circumnavigation in a few years; but the average temperature structure at a particular location remains the same.

We have seen that the distribution of oceanic temperatures is in part the direct result of insolation. Just as important are the processes of horizontal **advection** (horizontal movements) of air and water masses that transport warm air and water to cooler regions and vice versa. Figure 2.12 is a time-averaged map of the major surface current systems. In a similar generalized way, Figure 2.13 shows how the sinking of cold dense water in polar regions drives the circulation in the deep oceans. Ocean waters never stop moving, and this point is so crucial that it is worth repeating:

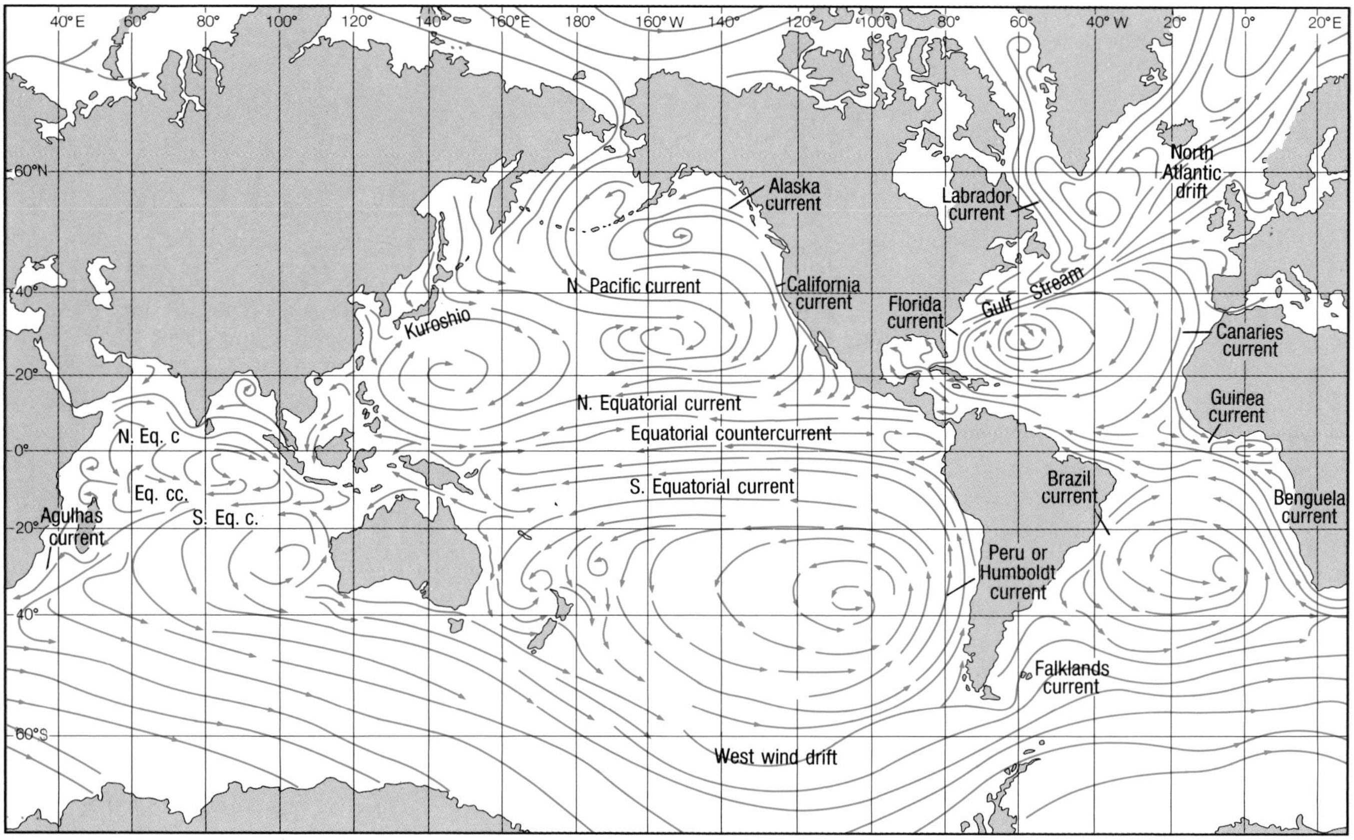

Figure 2.12 The world-wide average pattern of oceanic surface currents for November to April. There are significant differences in the Indian and western Pacific Oceans, due to seasonal reversals (monsoons), from May to October.

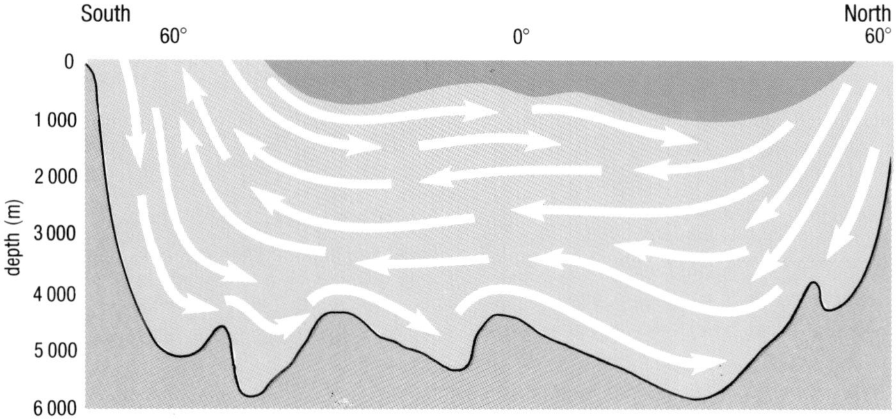

Figure 2.13 Diagrammatic section to illustrate the general form of the deep circulation in the Atlantic Ocean, driven by cold dense water sinking in high latitudes (the base of the shaded region at the top approximates to the 10 °C isotherm). Deep currents also circulate through the other major ocean basins.

the temperature (and, as you will see, the salinity) at any particular location and depth—at least below the mixed surface layer—will change hardly at all from year to year, even though the actual water at that location and depth is changing all the time.

2.6 SUMMARY OF CHAPTER 2

1 The Earth's surface temperature is mainly determined by the amount of insolation it receives. On average, about 70% of incoming solar radiation reaches the surface, directly or indirectly. The proportion varies with latitude, season and time of day, and the amount absorbed depends on the albedo of the surface. The oceans have a large thermal capacity because of the high specific and latent heat of water, and they act as a temperature buffer for the surface of the Earth as a whole. Annual insolation is greatest in low latitudes and least at the poles, mainly because of the angle that the Sun's rays make with the Earth's surface: the higher the latitude, the lower the angle.

2 Conduction, convection and especially evaporation/precipitation are the principal means by which heat and water are exchanged across the air–sea interface. The oceanic evaporation/precipitation cycle contributes about one-quarter of the global heat budget. Aerosol production at the sea-surface is another important mechanism for the transfer of water (and salts) into the atmosphere.

3 Insolation penetrates no more than a few hundred metres into the oceans, and most solar radiation is absorbed within the topmost 10m. Downward transfer of heat is mainly by mixing, as conduction is very slow (water is a very poor conductor of heat). The uppermost 200–300m are typically well mixed by winds, waves and currents (the mixed surface layer). Below the mixed layer in low and mid-latitudes lies the permanent thermocline, across which temperature declines to about 5°C, and below which temperatures decrease very gently to the bottom (typically between 0°C and 3°C). In mid-latitudes, a seasonal thermocline can develop during the summer, above the permanent thermocline. There may also be diurnal thermoclines, at depths of 10–15m.

4 The temperature difference across the thermocline can be utilized to generate electricity, using the principles upon which the domestic refrigerator is based. The main problem in this application is that of scale.

5 The long-term stability of the distribution of temperature within the ocean means that sections and profiles of average temperature do not change significantly from year to year. This stable thermal structure is maintained by the continuous three-dimensional motion of the global system of surface and deep currents.

Now try the following questions to consolidate your understanding of this Chapter.

QUESTION 2.6 (a) Explain why colder water can overlie water at 4°C in a freshwater lake in a gravitationally stable situation. Could such a situation develop in the oceans?

(b) Explain why the temperature profile of a freshwater lake will not show temperatures decreasing with depth to values of less than 4°C.

QUESTION 2.7 The broad thermal structure of the oceans allows us to recognize three main layers. Name them and summarize their characteristics.

CHAPTER 3 SALINITY IN THE OCEANS

The average concentration of dissolved salts in the oceans—the **salinity**—is about 3.5 per cent (%) by weight. Until the early 1980s, salinity values were expressed in parts per thousand, or per mil, for which the symbol is ‰; the average salinity quoted above is 35‰. It has become standard practice to dispense with the symbol because salinity is now defined in terms of a ratio, as explained in Section 3.3.3. Table 3.1 lists the 11 major elements that together make up 99.9% of the dissolved constituents of seawater. The symbol ‰ appears in Table 3.1 to remind you that the numbers represent concentrations in parts per thousand by weight. In accordance with modern practice, we shall normally dispense with this symbol hereafter and use only numbers when presenting salinity values.

In surface waters of the open oceans, salinity ranges from 33 to 37, but when shelf seas and local conditions are taken into account, the range can be as wide as 28–40 or more. **Brackish water** has a salinity of less than about 25, while **hypersaline** water has a salinity greater than about 40.

Table 3.1 Average concentrations of the principal ions in seawater, in parts per thousand by weight.

Ion	‰ by weight		
chloride, Cl^-	18.980	negative ions (anions) total	= 21.861‰
sulphate, SO_4^{2-}	2.649		
bicarbonate, HCO_3^-	0.140		
bromide, Br^-	0.065		
borate, $H_2BO_3^-$	0.026		
fluoride, F^-	0.001		
sodium, Na^+	10.556	positive ions (cations) total	= 12.621‰
magnesium, Mg^{2+}	1.272		
calcium, Ca^{2+}	0.400		
potassium, K^+	0.380		
strontium, Sr^{2+}	0.013		
		overall total salinity	= 34.482‰

Note: You will find tables of average concentrations of the dissolved constituents of seawater elsewhere, both in this Series and other literature. They may well differ in detail from Table 3.1 because different authorities use different sources to compile their averages.

Table 3.2 Approximate average percentages by weight of the ten most abundant elements (other than oxygen) in the Earth's crust.

Element	% by weight
Si	28.2
Al	8.2
Fe	5.6
Ca	4.2
Na	2.4
K	2.4
Mg	2.0
Ti	0.6
Mn	0.1
P	0.1

QUESTION 3.1 In Table 3.1, the proportion by weight of negative ions greatly exceeds that of positive ions. So why does seawater not carry a net negative charge?

Compare Table 3.1 with Table 3.2, which presents an approximate average elemental composition of crustal rocks: there are some obvious contrasts. These are particularly striking when you realize that the operation of the hydrological cycle provides most of the dissolved constituents in seawater; however, since the late 1970s oceanographers have recognized the important contribution to seawater composition made by **hydrothermal circulation** at ocean ridge crests.

QUESTION 3.2 How many of the major elements of common crustal rocks (Table 3.2) can you find in Table 3.1, and which are they?

The three most abundant elements in Table 3.2—let alone others—do not appear at all in Table 3.1. The reason lies in the degree of solubility and reactivity of different elements when rocks are weathered and the resulting products are carried away by rivers to the sea. Many of the commonest elements in rocks, such as silicon, aluminium and iron, are not very soluble, and so they are transported and deposited mainly in solid particles of sand and clay. Others such as sodium, calcium and potassium are relatively soluble and are transported mainly in solution. Hydrothermal solutions associated with sea-floor spreading supply some elements to seawater and remove others from it. The relative amounts of dissolved constituents within the oceans are controlled by complex chemical and biological reactions in seawater, as outlined in Chapter 6.

3.1 CONSTANCY OF COMPOSITION

The **constancy of composition of seawater** is an important concept in oceanography. For most of the major constituents in Table 3.1, the following generalization applies:

> The total amount of the major dissolved ions can vary from place to place in the oceans, but their relative proportions remain virtually constant.

In other words, the total salinity can change, but the ratio of the concentration of any particular major ion to the total remains virtually constant, and so do the ratios of the concentrations of individual major ions to one another.

QUESTION 3.3 (a) What is the ratio of potassium concentration to total salinity in Table 3.1?

(b) What would the potassium concentration be if the salinity (i) rose to 36, (ii) fell to 33?

(c) What is the ionic ratio of $K^+:Cl^-$ in Table 3.1? What would it be in each of cases (i) and (ii) of part (b)?

(d) How might these changes of salinity come about?

The way salinity varies throughout the oceans depends almost entirely on the balance between evaporation and precipitation and the extent of mixing between surface and deeper waters. In general, changes of salinity have no effect on the relative proportions of the major constituents—their concentrations all change in the same proportion, i.e. their ionic ratios remain constant.

Exceptions to this generalization are the small variations which have been detected in the proportions of calcium, bicarbonate* and magnesium. The first two are used by marine organisms to build skeletons of calcium

*The term 'bicarbonate' for HCO_3^- is being replaced by the term 'hydrogen carbonate' nowadays. However, we shall use the term 'bicarbonate' as it is still widely used in the literature.

carbonate, which may re-dissolve after the organisms die. The ratio of Ca^{2+} to total salinity is about 0.5% greater in deep than in surface waters, which is consistent with such a process (see Chapter 6). Magnesium can substitute for calcium in the skeletons of marine organisms. In addition, magnesium and calcium are important participants in the reactions between heated seawater and crustal rocks in sea-floor hydrothermal systems. Both processes may cause changes in the ionic ratios of magnesium and calcium.

3.1.1 CHANGES DUE TO LOCAL CONDITIONS

In some marine environments, away from the open oceans, conditions are such that ionic ratios show large departures from normal. Such regions include:

1 Enclosed seas, estuaries and other regions where there is a substantial inflow of river water, which not only contains significantly less total dissolved salts than seawater, but also has very different ionic ratios (see Chapter 6).

2 Basins, fjords and other regions where the bottom circulation is severely restricted, e.g. by the presence of a sill (a subsurface barrier) at the mouth of the basin, preventing free communication between the bottom water and the oxygenated oceanic water outside. In such cases, the bacterial breakdown (oxidation) of organic matter in the bottom water leads to depletion of dissolved oxygen, which may be severe enough to result in conditions that are described as **anoxic** or **anaerobic.** Sulphate anions are then used as an alternative source of oxygen by the micro-organisms.

3 Extensive areas of warm, shallow water, such as the Bahama Banks, which are characterized by very active chemical and/or biological precipitation of calcium carbonate, leading to changes in the ratio of Ca^{2+} to total salinity.

4 Areas of sea-floor where interstitial or pore waters in the sediments participate in a wide variety of reactions with the sediment particles during compaction, after the sediments have been deposited. Such reactions lead to considerable changes in ionic ratios.

5 Regions of sea-floor spreading and active submarine volcanism, where heated seawater circulates through cracks and fissures in the oceanic crust. Ionic ratios in hydrothermal solutions are very different from those of normal seawater, and the resulting mixed waters have quite atypical major element : salinity ratios.

QUESTION 3.4 As Table 3.1 shows, sulphur is present in seawater mainly as SO_4^{2-} and in practice it is measured in this form. Would you expect the ratio of SO_4^{2-} to total salinity to be higher or lower in anoxic basins (item 2 in the above list) than in open ocean waters?

3.1.2 SALTS FROM SEAWATER

As seawater evaporates, the least soluble salts reach saturation first, so the sequence of precipitation is in the order of increasing solubility, not of abundance. The sequence is shown in Figure 3.1, along with the relative proportions of the precipitated salts. The first to be precipitated is calcium carbonate ($CaCO_3$), which forms only a minute proportion of the salts because of the relatively low abundance of bicarbonate ions (Table 3.1).

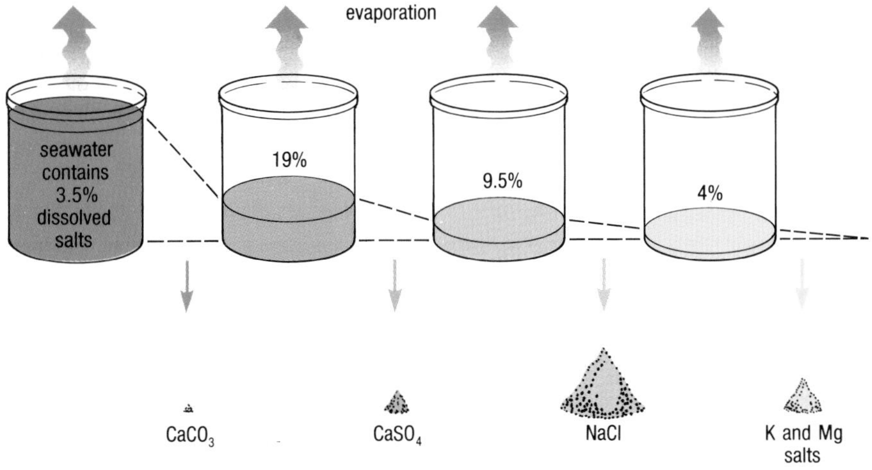

Figure 3.1 The succession of salts precipitated from seawater. On evaporation, $CaCO_3$ is precipitated first. When evaporation has reduced the volume to 19% of the original amount, $CaSO_4$ begins to precipitate; at 9.5% of the original volume, NaCl starts to precipitate, and so on. The volumes of the piles represent the relative proportions of the precipitated salts.

Calcium sulphate is precipitated either as anhydrite ($CaSO_4$) or as gypsum ($CaSO_4.2H_2O$), depending upon the conditions. Sodium chloride (halite, NaCl) is the most abundant salt, and the residual brine contains the chlorides of potassium and magnesium, which are the most soluble and therefore the last to be precipitated.

Virtually every coastal nation has at some time produced sea salt commercially and about 60 countries still do, either by industrial processes or by solar evaporation (Figure 3.2). Some 40 million tonnes of sodium chloride are extracted from seawater each year world-wide, some for human consumption but most of it for manufacturing chemicals. Magnesium hydroxide is chemically precipitated from seawater and used to produce around 750000 tonnes of magnesium and its compounds annually. Bromine is released by electrolysis as a gas and then condensed to liquid—annual production is about 40000 tonnes.

These are among the major constituents of seawater, but most elements dissolved in seawater occur in minute concentrations (see Chapter 6). The total volume of seawater is so huge, however, that tonnages are enormous and efforts to extract valuable elements such as gold and uranium have been made many times; but no method has so far proved economic.

(a)

(b)

Figure 3.2 (a) Raking sea salt out of the residual brine in solar evaporation pans near Aveiro in Portugal.

(b) Carrying salt from the piles to lorries or boats for transport to the purification and processing centre.

3.2 VARIATIONS OF SALINITY

The distributions of temperature and salinity together provide oceanographers with information that enables them to trace the three-dimensional pattern of oceanic circulation. This Section describes how salinity varies vertically and horizontally in the oceans. As in the case of temperature distribution, the maps and profiles illustrate a long-term stability of salinity distribution which is maintained dynamically. As you read, bear in mind that the salinity at any particular location may hardly change from year to year, but the water at that location is changing all the time (*cf.* Section 2.5).

3.2.1 DISTRIBUTION OF SALINITY WITH DEPTH

Figure 3.3 shows a vertical section illustrating the relatively restricted range of salinity encountered in the main body of the oceans. Surface salinities depend mainly on the balance between precipitation and evaporation, which is climatically controlled (*cf.* Question 3.3(d)). Below

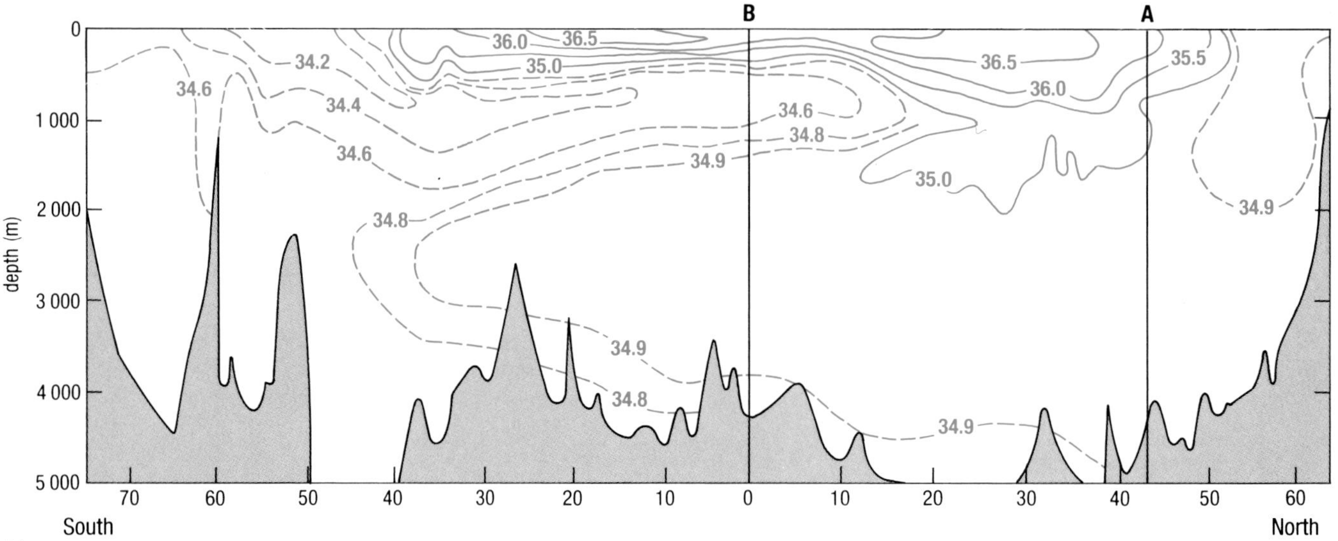

(a)

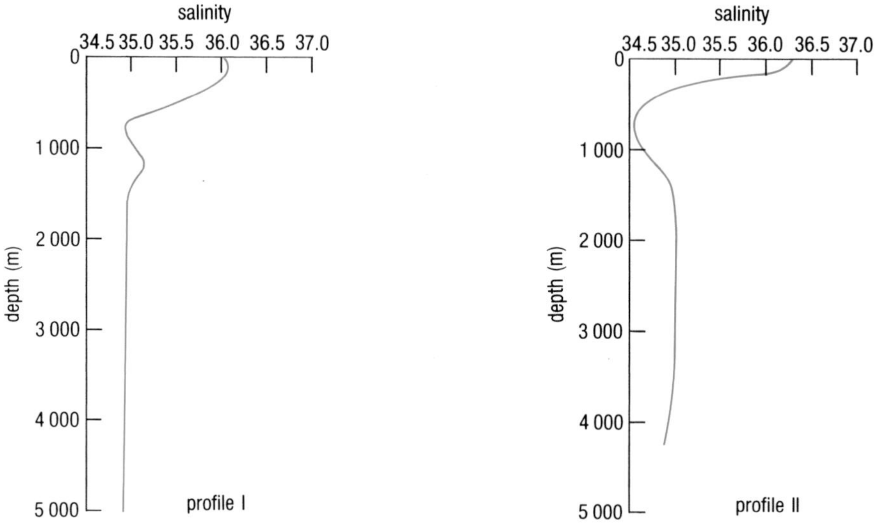

(b)

Figure 3.3 (a) A vertical section showing the distribution of salinity in the western Atlantic Ocean which illustrates that the range of salinity in surface layers is much greater than that of the main body of ocean water below 1 000 m. This general pattern is typical of all ocean basins, although the detail will vary from ocean to ocean. Note the great vertical exaggeration. Lines joining points of equal salinity are called **isohalines**. Broken lines 0.1 and 0.2 interval; solid lines 0.5 interval. The vertical lines A and B relate to (b) and are for use with Question 3.5.

(b) Salinity profiles along lines A and B in (a), for use with Question 3.5.

about 1000m, however, the influence of surface fluctuations is negligible and salinities are consistently between about 34.5 and 35 at all latitudes.

Zones where salinity decreases with depth are typically found at low and middle latitudes, between the mixed surface layer and the top of the deep layer, in which the salinity is roughly constant. These zones are known as **haloclines.**

QUESTION 3.5 (a) Which salinity–depth profile in Figure 3.3(b) corresponds to which vertical line in Figure 3.3(a)? How do the depth ranges of the haloclines compare with those of the thermoclines shown earlier in Figure 2.6?

(b) Which of the two haloclines in Figure 3.3(b) shows the greater rate of decrease of salinity with depth?

(c) Would you expect salinity–depth profiles for high latitudes to be similar to those in Figure 3.3(b)? Is there a halocline in those regions?

3.2.2 DISTRIBUTION OF SURFACE SALINITY

The salinity of the surface waters of the oceans is at a maximum in latitudes of about 20°, where evaporation exceeds precipitation. These regions correspond to the hot barren deserts that exist in similar latitudes on land. Salinities decrease both towards higher latitudes and towards the Equator (Figure 3.4). Local modifications are superimposed on this regional pattern, particularly near land masses. Surface salinity may be reduced by an influx of freshwater at the mouths of large rivers, and by melting ice and snow at high latitudes. On the other hand, surface salinities tend to be high in lagoons and other partly enclosed shallow marine basins at low latitudes, where evaporation is high and the inflow of water from adjacent land areas is limited.

QUESTION 3.6 Examine the map (a) and profiles (b) in Figure 3.4, and explain the maxima and minima on the profiles; then account for the salinity minimum in equatorial latitudes.

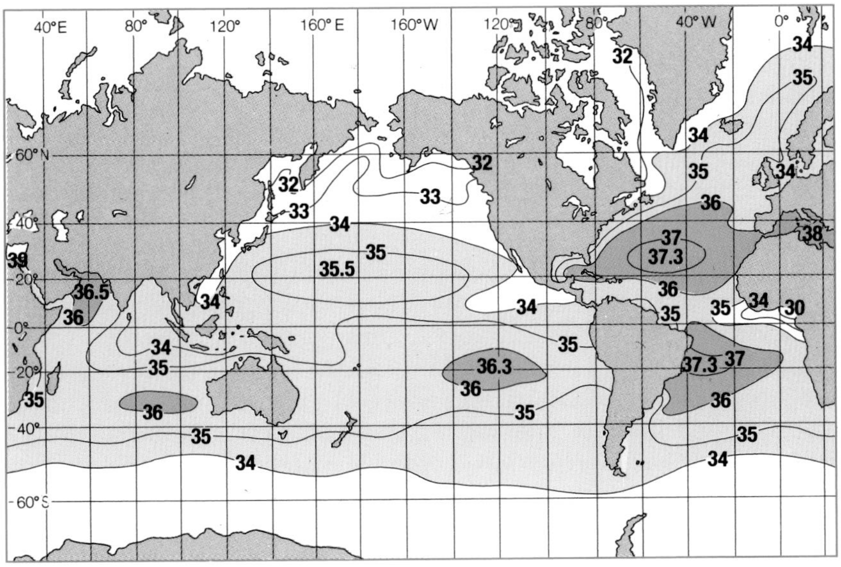

(a)

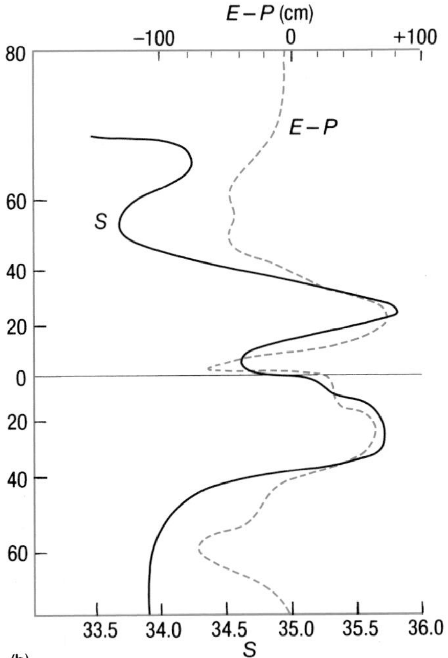

(b)

Figure 3.4 (a) The approximate positions of mean annual isohalines.

(b) Average values of salinity, S (black line) and the difference between average annual evaporation and precipitation ($E - P$) (blue line), plotted against latitude.

3.3 THE MEASUREMENT OF SALINITY

Early attempts to determine the chemical composition of seawater were hampered by the low sensitivity of analytical techniques. It was not until the early nineteenth century that any order became apparent in the data and the constancy of composition of seawater was first recognized from the few analyses available. During the cruise of HMS *Challenger* from 1872 to 1876, 77 water samples were collected from various depths in nearly all the major oceans and seas, and analysed for the elements chlorine, sodium, magnesium, sulphur, calcium, potassium and bromine. The method used for each element was rigorously tested on synthetic samples, thus giving a check on the reliability of the technique.

Since the nineteenth century, a large number of investigations have been carried out into the ratio of single constituents to salinity. During the mid-1960s, scientists from the British National Institute of Oceanography (now the Institute of Oceanographic Sciences) and the University of Liverpool analysed more than one hundred samples for all the major constituents. In the 1970s, the GEOSECS programme (GEochemical Ocean SECtionS), based in the USA, collected systematic chemical data for all the oceans, using the most accurate analytical techniques available and (more importantly) sampling procedures that minimized contamination. The huge amount of data collected is still being interpreted.

3.3.1 CHEMICAL METHODS OF SALINITY MEASUREMENT

Perhaps the most obvious way of measuring salinity is to take a known amount of seawater, evaporate it to dryness and then weigh the remaining salts (gravimetric determination). Although simple in theory, such a method gives variable and therefore unreliable results, for a number of reasons. The residue left after evaporation is a complex mixture of salts, together with some water of hydration bound to the solids, plus a small amount of organic material. The amount of water left behind can obviously be decreased by thorough drying of the residual salts at elevated temperatures, but this leads to other problems such as (i) decomposition of some of the salts (e.g. loss of HCl from hydrous $MgCl_2$ crystals); (ii) vaporization and decomposition of the organic matter; and (iii) expulsion of CO_2 from carbonate salts. The weight of solid material left behind after evaporation (and hence the measured salinity) thus depends on the conditions employed to drive off the water. Marine chemists in the nineteenth century were well aware of this in their attempts to measure salinity gravimetrically, and devised procedures that gave reasonably reproducible results.

None the less, gravimetric determination of salinity remains both difficult and tedious, and so other methods have been investigated. As you have read in Section 3.1, the concentrations of many major dissolved constituents of seawater bear a constant ratio to the total dissolved salt concentration, so the concentrations of one or more major constituents can be used to deduce the total salinity, S. The easiest constituents to measure are the halides (chloride + bromide + iodide), and this led ultimately to the empirical relationship:

$$S = 1.80655\ Cl \qquad (3.1)$$

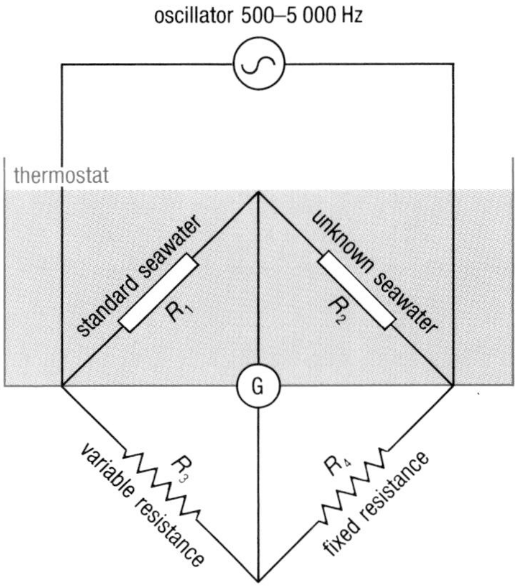

Figure 3.5 A simple bridge circuit for measuring conductivity of seawater samples. R_1 to R_4 represent resistance values, and G is the galvanometer. The thermostat would commonly be a constant-temperature water bath.

where *Cl* is the **chlorinity** of the sample, defined as the concentration of chloride in seawater (in parts per thousand) assuming that the bromide and iodide have been replaced by chloride.

Chlorinity was measured by titration and the salinity determined by substitution in equation 3.1 or tables derived from it. This method was used for determining virtually all salinities from the turn of the century until the mid-1960s. It is rarely used today, having been almost entirely superseded by measurements of electrical conductivity.

3.3.2 PHYSICAL METHODS OF SALINITY MEASUREMENT

Pure water is a poor conductor of electricity. However, the presence of ions in water enables it to carry an electric current. In the 1930s it was established that the electrical **conductivity** of seawater is proportional to its salinity. Conductivity is inversely proportional to resistance.

The first conductivity salinometers employed simple circuits of the type shown in Figure 3.5. In operation, R_3 is adjusted until the current flowing through the galvanometer falls to zero, when the following relationship holds:

$$\frac{R_1}{R_2} = \frac{R_3}{R_4} \tag{3.2}$$

As the values of all the resistances, except R_2, are known, the resistance of the seawater sample can be calculated. For oceanographic work, water of salinity of 35 was generally used as the standard water (R_1) for calibration purposes.

Conductivity is also affected by the temperature of the solution, which can lead to appreciable errors. Physical oceanographers ideally require salinity measurements to be accurate to ± 0.001 if possible. To achieve this, conductivity must be measured to 1 part in 40000. A change of this magnitude can be induced by a temperature change of only 0.001°C, so careful control of temperature is essential. Precision thermostating can maintain both sample and standard seawater at the same temperature, but the equipment is then bulky and measurements take a long time, because the sample has to be heated or cooled to working temperature before measurement can begin. Such problems inherent in circuits of the type shown in Figure 3.5 have now been largely circumvented, and modern salinometers are compact and rapid in operation, and can measure salinity to $\pm 0.003^0/_{00}$ or better. Conductivity sensors have been incorporated into *in situ* temperature–salinity instruments for use in shallow waters, and into conductivity–temperature–depth (CTD) probes for use in the deep oceans.

3.3.3 THE FORMAL DEFINITION OF SALINITY

Since the mid-1960s, the definition of salinity has been based (by international agreement) on empirically determined and rather complicated-looking formulations involving a conductivity standard. The formal definition that has been in use since the early 1980s runs as follows.

The practical salinity of a sample of seawater is defined in terms of the conductivity ratio, K_{15}, which is defined by:

$$K_{15} = \frac{\text{conductivity of seawater sample}}{\text{conductivity of standard KCl solution}} \tag{3.3}$$

at 15°C and 1 atmosphere pressure, the concentration of the standard KCl solution being 32.4356g kg^{-1}.

The practical salinity is related to the ratio K_{15} by the following equation:

$$S = 0.0080 - 0.1692\, K_{15}^{1/2} + 25.3851\, K_{15} + 14.0941\, K_{15}^{3/2} - 7.0261\, K_{15}^{2} + 2.7081\, K_{15}^{5/2} \tag{3.4}$$

1 You do not need to remember any details of equation 3.4. Question 3.7 (below) is intended simply to show you how it works.

2 As the definition is a ratio, salinities are nowadays presented simply as numbers. It is important to remember that the numbers represent grammes per kilogramme, or parts per thousand by weight.

3 In practice, tables (and, increasingly, computer algorithms) are used for the direct conversion of K_{15} into S, and for converting conductivity ratios at temperatures and pressures of measurement other than 15°C and 1 atmosphere pressure into K_{15}.

QUESTION 3.7 Use equation 3.4 to answer this question by completing the following sentence. By definition, when $K_{15} = 1$, the practical salinity is exactly equal to...?

A salinity value determined by conductivity depends on the temperature and pressure at which the conductivity is measured, and is thus somewhat removed from the simple but fundamental idea of salinity being the total dissolved salts in a seawater sample. In fact, for open ocean seawater, the two are closely related: the content of total dissolved salts in g per kg of seawater is $1.00510 \times S$, where S is as defined in equation 3.4.

3.4 SUMMARY OF CHAPTER 3

1 The average salinity of seawater is close to 35 parts per thousand (‰) by weight. Eleven ions make up 99.9 per cent of the dissolved constituents: Cl^-, Na^+, SO_4^{2-}, Mg^{2+}, Ca^{2+}, K^+, HCO_3^-, Br^+, $H_2BO_3^-$, Sr^{2+} and F^-, in that order. The relative proportions of elements in solution in seawater differ greatly from the proportions in crustal rocks, because of their different solubilities in the solutions formed during terrestrial weathering and sea-floor hydrothermal activity.

2 Salinity varies from place to place in the oceans, but the relative proportions of most major dissolved constituents (their ionic ratios) remain virtually constant. Evaporation and precipitation change the total salinity, but do not affect the constancy of composition.

3 Minor departures from constancy of composition in the open oceans result mainly from the intervention of biological and hydrothermal processes, affecting principally Ca^{2+}, Mg^{2+} and HCO_3^-. Major departures are the result of local conditions, chiefly in shallow near-shore waters, but also where hydrothermal activity occurs. Some major constituents are extracted commercially from seawater, but minor constituents have so far not been successfully extracted on a commercial scale.

4 The vertical and lateral distribution of salinity in the oceans do not change significantly from year to year, but the waters themselves are continually moving in a three-dimensional system of surface and deep currents. Surface salinities in the open oceans are greatest (up to 38) in tropical latitudes, where evaporation exceeds precipitation. They are lower near the Equator (*c.* 35) and in high latitudes (*c.* 33–34), because of greater rainfall and (in high latitudes) melting ice and snowfall. In middle and low latitudes, there is a halocline from the base of the mixed surface layer to about 1000m depth, below which salinities are generally between 34.5 and 35.

5 Gravimetric measurement of salinity is difficult because of decomposition of some of the salts on heating to evaporation. Chemical measurements of salinity, based on titration to determine chlorinity, were standard until the 1960s, but have been almost entirely superseded by electrical conductivity methods. An empirically determined formula is used to convert conductivities, measured against a standard, into salinities.

Now try the following questions to consolidate your understanding of this Chapter.

QUESTION 3.8 Which of the following statements (a)–(e) are true, and which are false?

(a) The relative proportions of elements dissolved in seawater are very similar to those in average crustal rocks.

(b) Salinity can vary from place to place in the oceans, but the ratio of salinity to chlorinity will nearly always remain constant.

(c) The ratio of Ca^{2+} to salinity will fall where there is significant precipitation of calcium carbonate.

(d) Haloclines are regions in which salinity increases with depth.

(e) It is impossible to measure either salinity or temperature to better than one part in a thousand.

QUESTION 3.9 The oceans are not a closed system, and large amounts of dissolved salts are continually being introduced into them from the world's rivers. There are also significant inputs from other sources such as hydrothermal solutions. So how is it that, in general, the constancy of composition of seawater is maintained?

CHAPTER 4 DENSITY AND PRESSURE IN THE OCEANS

The vertical and horizontal distributions of isotherms and isohalines remain fairly constant from year to year; relatively small seasonal fluctuations are confined to the surface layer. We have emphasized that these distributions represent a form of *dynamic equilibrium* or steady state, because the ocean waters themselves are continuously moving. The motion is not random, but is organized in a three-dimensional circulation system that shows little variation when motions are averaged out over periods of several years.

4.1 WATER MASSES

The Earth's climate and weather are largely the result of the movements of large air masses, each characterized by particular and recognizable combinations of temperature, humidity and pressure. In much the same way, large **water masses** in the ocean move vertically and horizontally, each defined by its temperature (T), salinity (S) and other characteristics, which can be used to identify it and track its movements. The main features of the movement of water masses are summarized below:

1 Figure 4.1 shows the boundaries of water masses formed in upper parts of the oceans, extending from surface or near-surface waters down to about the base of the thermocline. They are identified by their temperature, salinity and other properties, including the communities of organisms that inhabit them. If you compare Figure 4.1 with Figure 2.12, you can see that the boundaries between these upper water masses coincide quite well with major surface current systems. It is also possible to identify boundaries between water masses moving in different directions at greater depths in the oceans.

QUESTION 4.1 (a) Look at Figure 2.13 and try to sketch in the boundaries between three water masses.

(b) Water masses can be identified by their temperature (T) and salinity (S) signatures. In what way would you expect these properties to change (i) within and (ii) at the boundaries of water masses?

2 Water moves much more slowly than air, so water masses are less variable than air masses, and their boundaries do not change much, even on time-scales of decades to centuries.

3 The surface current systems are driven by winds, but the movement of intermediate and deep water masses is controlled by density. When the density of the surface layers of seawater is sufficiently increased, the water column becomes gravitationally unstable and the denser waters sink.

QUESTION 4.2 (a) How can the density of surface waters in the oceans be increased in (i) polar and (ii) tropical regions?

(b) Is it reasonable to regard density-driven circulation in the ocean depths as a consequence of interactions between the atmosphere and oceans?

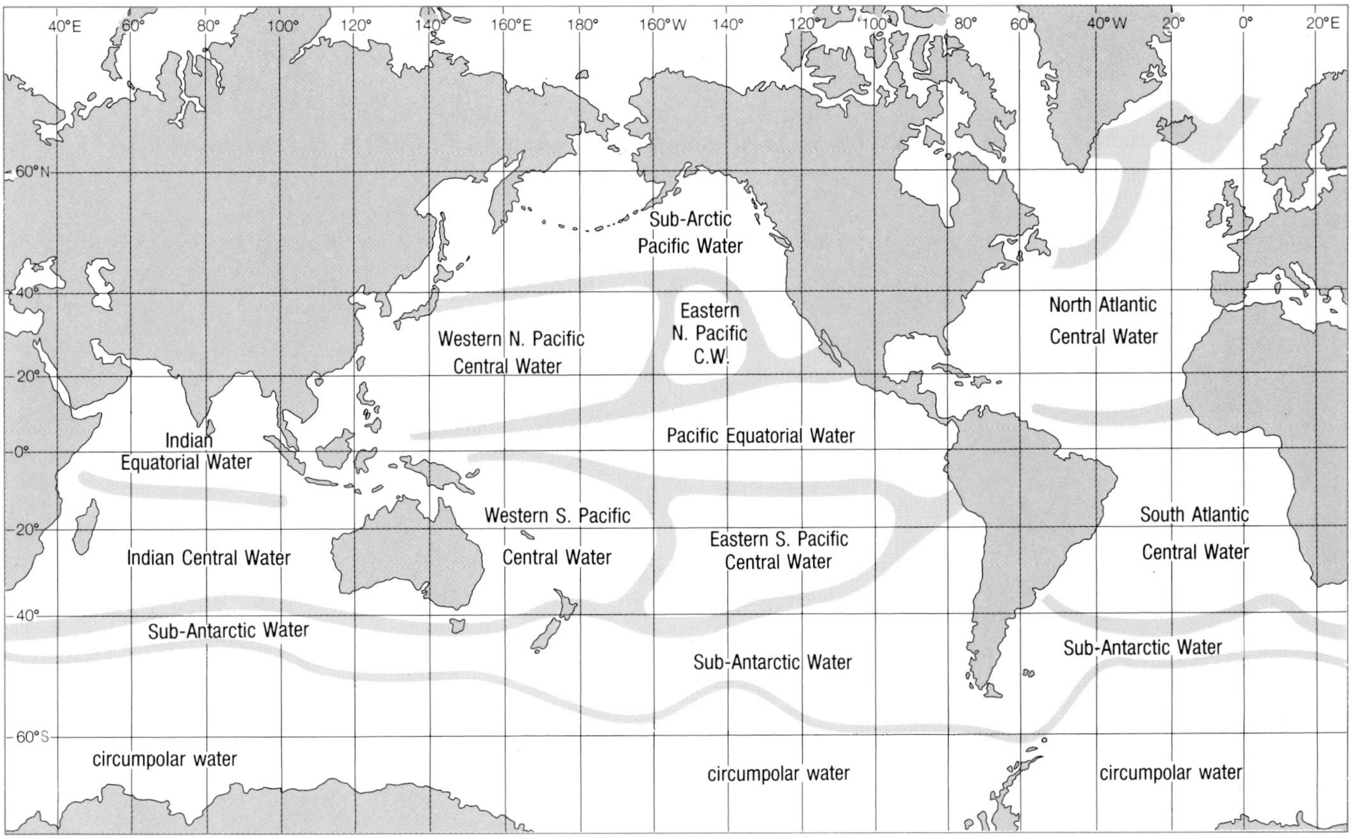

Figure 4.1 The approximate boundaries of the upper water masses of the oceans (*cf.* Figure 2.12).

Vertical circulation in the oceans is controlled by variations in both temperature and salinity, and it is known as the **thermohaline circulation**. Its principal components are the cold dense water masses produced around the polar ice-caps, which sink to the sea-floor and thence spread throughout the oceans, flowing beneath all the other water masses. The water from the Antarctic (Antarctic Bottom Water (AABW, Figure A1)) actually crosses the Equator into the Northern Hemisphere. In the North Atlantic, there are comparable southward-flowing bottom currents coming from the Arctic, but there are no such currents in the North Pacific because of the barrier formed by the Aleutian island arc to the north.

In the surface waters of the oceans, temperature and salinity alone control the density of seawater, but in the deep oceans another factor becomes important: pressure.

4.2 DEPTH (PRESSURE), DENSITY AND TEMPERATURE

The density of seawater does vary somewhat with depth, but not to the extent supposed barely a century-and-a-half ago (Figure 4.2):

> 'The enormous pressure at these great depths seemed at first sight alone sufficient to put any idea of life out of the question. There was a curious popular notion, in which I well remember sharing when a boy, that, in going down, the seawater became gradually under the pressure heavier and heavier, and that all the loose things in the sea

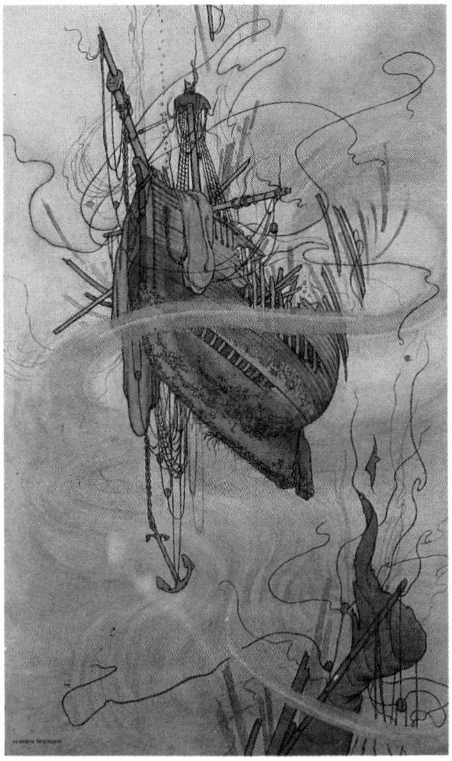

Figure 4.2
'The wrecks dissolve above us;
their dust drops down from afar –
Down to the dark, to the utter dark . . . .'
Rudyard Kipling, 'Song of the English'.

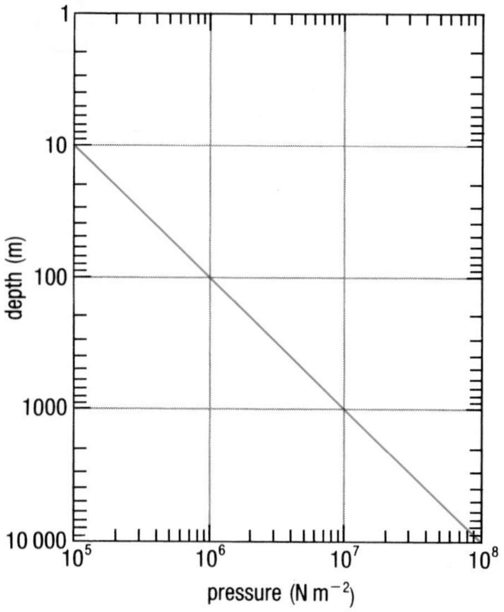

Figure 4.3 Graph of pressure against depth in the oceans. Both scales are logarithmic simply to accommodate the range of numbers. The relationship between pressure and depth is linear. (Pressure is measured in newtons per square metre; 10^5 N m^{-2} = 1 bar ≈ 1 atmosphere.)

floated at different levels, according to their specific weight: skeletons of men, anchors and shot and cannon, and last of all the broad gold pieces wrecked in the loss of many a galleon on the Spanish Main; the whole forming a kind of "false bottom" to the ocean, beneath which there lay all the depth of clear still water, which was heavier than molten gold.'

C. Wyville Thompson (1873) *The Depths of the Sea*, Macmillan.

The effect of pressure on density is not quite so dramatic as that. However, we should not scoff at such notions. Indeed, the concept of neutral buoyancy which is implicit in them is applied in modern technology (Section 5.2.3).

The **hydrostatic equation** describes the way in which pressure P is related to depth (or height) (z) in a column of fluid.

$$P = g\rho z \qquad (4.1)$$

where g is the acceleration due to gravity and ρ (rho) is the density.

Provided density remains constant, the hydrostatic equation shows a proportional relationship between pressure and depth (height). It is generally valid for the oceans, because water is only slightly compressible and the density of 99% of seawater is within ± 2% of its mean value of about 1.03×10^3kgm^{-3}. On the scale of Figure 4.3, the result is a straight line.

QUESTION 4.3 (a) The value of g is 9.8ms^{-2}. Use the hydrostatic equation to work out the pressure due to a 10m column of seawater. Your answer will be in Nm^{-2} (newtons per square metre). How does your answer compare with the value for normal atmospheric pressure?

(b) Approximately what pressures characterize (i) most of the deep ocean floors and (ii) the ocean trenches?

4.2.1 ADIABATIC CHANGES

Adiabatic changes are those that occur independently of any transfer of heat to or from the surroundings. They are a consequence of the compressibility of fluids. When a fluid expands, it loses internal energy and its temperature falls. When compressed, it gains internal energy and its temperature rises—that is the principal reason why the pump heats up when you inflate your bicycle tyres. The principles of adiabatic gain and loss of heat on compression and expansion of gases provide the basis of refrigeration and air conditioning technology. As air rises into a region of lower pressure it expands, and the rate of fall of temperature for dry air is 9.8°Ckm^{-1}. Liquids are much less compressible than gases, and the rate of change of temperature with depth in the oceans as a result of adiabatic changes is less than 0.2°Ckm^{-1}.

This brings us to the very important concept of **potential temperature**. In both the oceans and the atmosphere it is defined as the temperature which the fluid would attain if brought adiabatically from the pressure appropriate to its actual height or depth to a pressure of 1000 millibars (i.e. approximately one atmosphere at sea-level). It is thus different from the ***in situ* temperature**, which is the temperature of the fluid measured at its actual height or depth.

Because of the great contrast in compressibility, the difference between potential and *in situ* temperature may be tens of degrees in the atmosphere, but is never more than about 1.5°C in the oceans. The latter figure may seem to be trivial, but you will soon see that potential temperature is a very important concept when we come to consider vertical temperature distribution and gravitational stability in the oceans.

QUESTION 4.4 Explain whether you would expect the potential temperature of (a) air at a height of 5 km and (b) seawater at a depth of 5 km to be greater or less than their respective *in situ* temperatures.

4.3 *T–S* DIAGRAMS

T–S* diagrams** are used to plot temperature and salinity data for water samples, and hence to identify water masses. Figure 4.4 is a *T–S* diagram. The contours represent density. The numbers are values of σ_t (**sigma-*t), which is a concept widely used in physical oceanography.

4.3.1 THE CONCEPT OF σ_t

σ_t is a shorthand way of expressing the density of a sample of seawater at *atmospheric pressure*, as determined from its *in situ* temperature and salinity. The simplest way of explaining the relationship of σ_t to density is to give an example. On Figure 4.4, the σ_t of seawater at 5°C (*in situ*

Figure 4.4 A *T–S* diagram contoured in σ_t values.

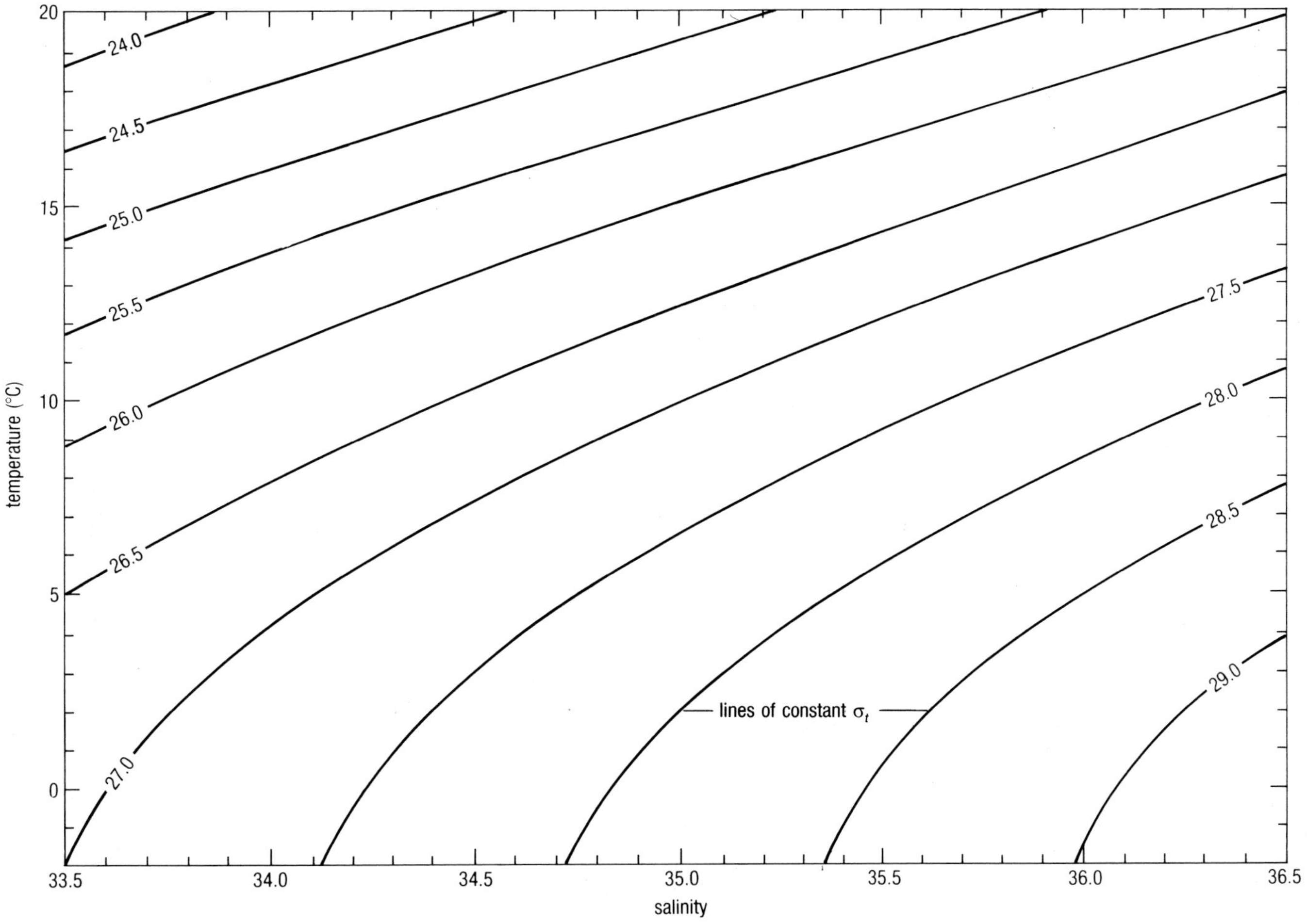

temperature) and salinity 33.5 is 26.5. The specific gravity of that water is 1.0265, and its density is $1.0265 \times 10^3 kg m^{-3}$. In general terms, then:

$$\sigma_t = (\text{specific gravity} - 1) \times 1000 \qquad (4.2)$$

Note that because specific gravity is used rather than density, σ_t has no units.

QUESTION 4.5 To make sure you understand the concept of σ_t, attempt these simple questions.

(a) What are the values of σ_t for seawater at (i) a temperature of 2°C and salinity of 34.5, and (ii) a temperature of 15°C and salinity of 35.6?

(b) What do those σ_t values mean in terms of (i) specific gravity at atmospheric pressure and (ii) density at atmospheric pressure?

(c) By looking carefully at Figure 4.4, and bearing in mind the range of T and S normally found in the oceans, would you say that, in general, temperature or salinity is likely to have the greater effect on the density of seawater?

We have seen that below depths of about 500–1000m in the oceans, temperature and salinity do not vary much. Figure 4.5(a) shows how this is reflected in the rather small increase of σ_t with depth below about 1000m. The curve of σ_t against depth is almost vertical below about 2000m.

In contrast, at depths of less than 500m in middle and low latitudes σ_t increases very rapidly with depth, and the curves in Figure 4.5(a) are almost horizontal. A marked step in the density profile is termed a **pycnocline**. In the open oceans, pycnoclines are usually associated with thermoclines, though their exact position and slope will also depend upon the distribution of salinity. The main pycnocline coincides approximately with the permanent thermocline. Water in a pycnocline is necessarily very stable, i.e. it takes a large amount of energy to displace it up or down. The main pycnocline forms a lower limit or 'floor' to turbulence caused by mixing processes at the surface.

The depth of the mixed surface layer (Section 2.3) depends on the strength of the wind and on the processes that tend to promote vertical stability, such as heating of the surface and precipitation.

How do heating and precipitation promote stability?

Both reduce the density of surface waters: warm water is less dense than cold water and freshwater is less dense than seawater. The very process of mixing in this upper layer, however, has the effect of increasing stability at its base, for that is how the pycnocline develops (Figure 4.5(b)).

4.3.2 THE CONCEPT OF σ_θ AND VERTICAL STABILITY

Although T–S diagrams are extremely useful for identifying and tracing water masses in the oceans, they can give a spurious impression of gravitational instability in deep waters.

Bearing in mind what you read earlier about potential temperature and how σ_t is determined, can you see how this could happen?

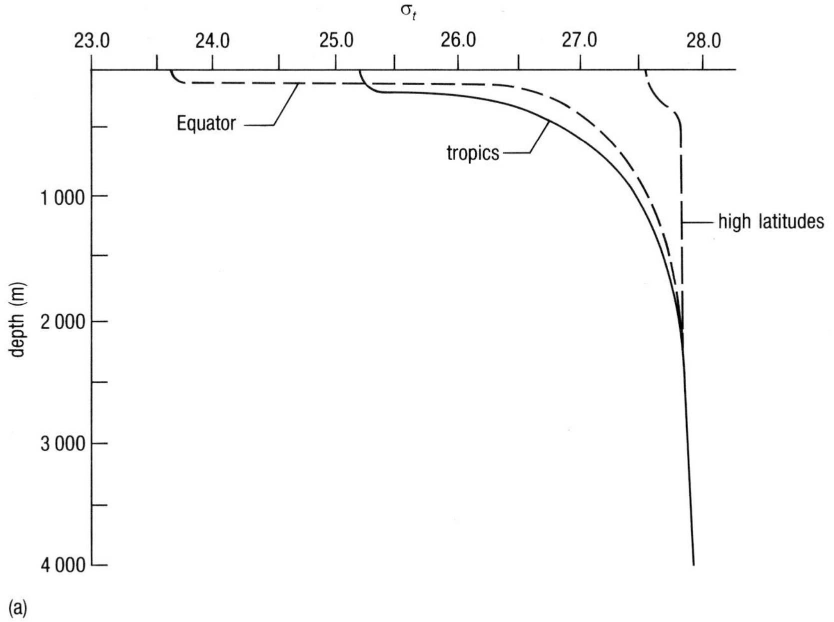

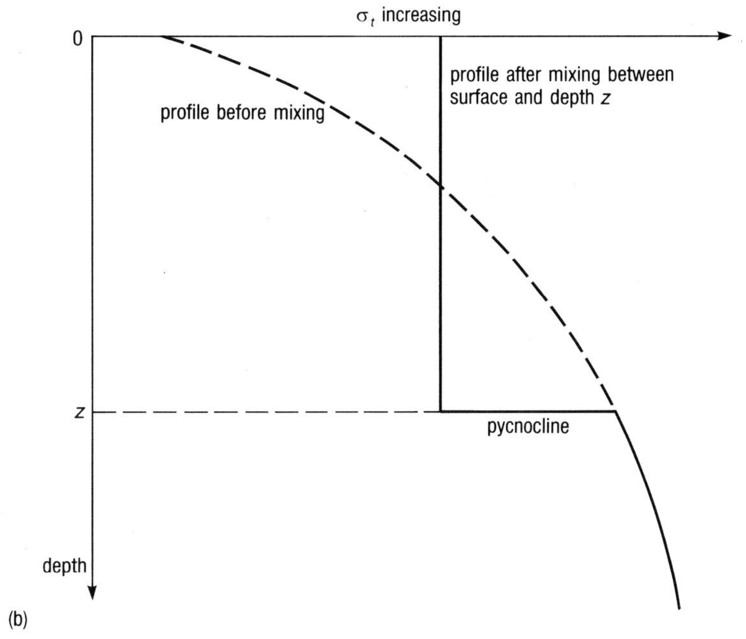

Figure 4.5 (a) Depth profiles of σ_t for different latitudes. The almost vertical part below about 2 000 m, where all three curves coincide, results from the very small regional variations of temperature and salinity in the deep oceans. Regions in which density changes sharply with depth are known as pycnoclines.

(b) Formation of the mixed surface layer changes the density–depth profile, resulting in the development of a pycnocline at the base of the layer. Note the very different scale compared to (a).

Density must increase with depth, to ensure gravitational stability in the oceans. Adiabatic compression raises the temperature of deep water, so that *in situ* temperature becomes progressively greater than its potential temperature with increasing depth. But σ_t is determined using *in situ* temperature uncorrected for adiabatic changes, so it will represent a density lower than that actually possessed by the water at any particular depth. In some cases, the differences are small enough to ignore, but it can happen that plots of salinity and *in situ* temperature show a decrease of σ_t with increasing depth, especially for deep water samples. These instablities disappear when the potential temperature, θ (Section 4.2.1) is used with salinity on a *T–S* diagram to determine the **potential density**, which is represented by the symbol σ_θ (**sigma-theta**).

Recalling that the potential temperature is defined as the temperature which a sample of water would have if brought adiabatically from depth to atmospheric pressure, would you expect σ_t and σ_θ for surface-water samples to be the same?

As water at the surface is under atmospheric pressure, there is no need to make the adiabatic correction, so σ_t and σ_θ for surface-water samples must be the same.

Table 4.1 shows how, in the Mindanao Trench off the Philippines, σ_t (calculated from observed salinity and *in situ* temperature measurements) increases down to 4450m and then decreases again. This suggests that the

Table 4.1 Comparison of *in situ* and potential temperatures in the Mindanao Trench off the Philippine Islands.

Depth (m)	Salinity	Temperature *in situ* (°C)	Temperature potential (°C)	Density σ_t	Density potential (σ_θ)
1455	34.58	3.20	3.09	27.55	27.56
2470	34.64	1.82	1.65	27.72	27.73
3470	34.67	1.59	1.31	27.76	27.78
4450	34.67	1.65	1.25	27.76	27.78
6450	34.67	1.93	1.25	27.74	27.79
8450	34.69	2.23	1.22	27.72	27.79
10035	34.67	2.48	1.16	27.69	27.79

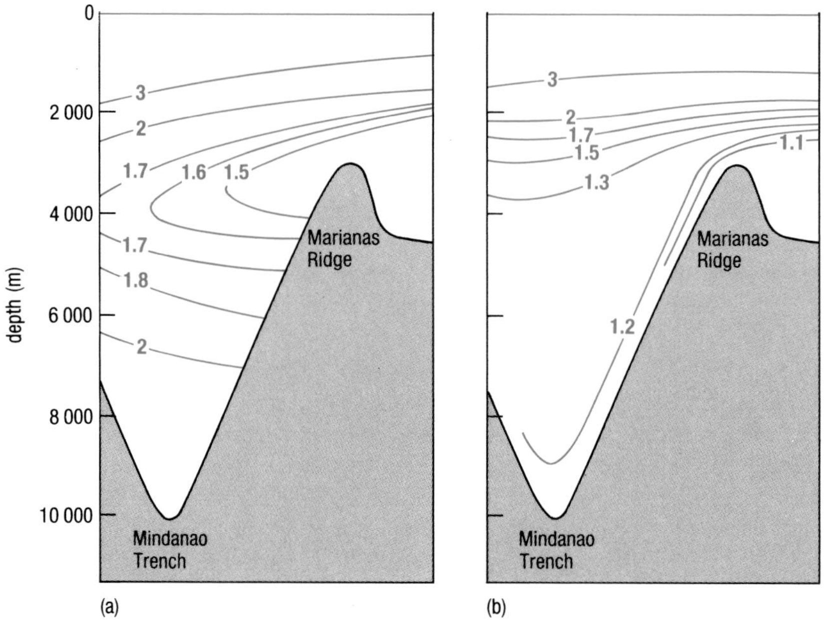

Figure 4.6 Two patterns of temperature distribution in the Mindanao Trench (for use with Question 4.6). Contours are in °C and represent either *in situ* temperature, or potential temperature, θ.

water column is gravitationally unstable. When *in situ* temperatures are corrected to potential temperatures, σ_t is replaced by σ_θ and the instability disappears.

QUESTION 4.6 Figure 4.6 summarizes part of Table 4.1 in diagrammatic form.

(a) Match each of the diagrams with one of the following descriptions: (i) cool water flows over the sill (Marianas Ridge), descending slightly across the Mindanao Trench but remaining in the middle depths to leave warmer bottom water undisturbed; (ii) cool water passes over the sill and flows down the slope to the bottom of the trench.

(b) Assuming that salinity is not exerting any significant control over density, which of the two situations, (i) or (ii), described in (a) is likely to be the real state of affairs?

Table 4.1 shows that the difference between *in situ* and potential temperature exceeds 1°C below depths of 8km, while even at depths of about 1km it is significant to the first decimal place. The differences clearly become much smaller as depth decreases, but it is important to recognize that there are small adiabatic temperature gradients even in the mixed surface layer, which is otherwise isothermal. The differences may be small, but the sensitivity of modern instruments means that in some circumstances it is worth making the correction for adiabatic effects even in the top 200m of the oceans.

4.3.3 THE USE OF *T–S* DIAGRAMS

In Section 4.1 you read that water masses can be identified by their *T–S* 'signatures'. For example, the three major subsurface water masses in the Atlantic Ocean (whose boundaries are approximately delineated in Figure A1) are characterized by the following narrow ranges of temperature and salinity:

Antarctic Bottom Water (AABW)	−0.5° to 0°C and 34.6 to 34.7
North Atlantic Deep Water (NADW)	2° to 4°C and 34.9 to 35.0
Antarctic Intermediate Water (AAIW)	3° to 4°C and 34.2 to 34.3

T–S diagrams can thus be used both to identify water masses and to determine the extent to which they have mixed with one another. For example, Figure 4.7 is a *T–S* diagram on which *T* and *S* data for a station in the southern equatorial Atlantic have been plotted. The *T* and *S* 'signatures' of the three water masses quoted above are also shown.

The water between about 1400m and 3800m depth represents the NADW, scarcely modified at all by mixing, even at these low latitudes (the station represented in Figure 4.7 lies south of the Equator). For simplicity, we are treating the NADW as a single water mass, but in fact it comprises more than one, with source regions in the Norwegian and Greenland and Labrador Seas.

The influence of the AABW is identifiable at the bottom of the profile in Figure 4.7, even though this bottom water has travelled thousands of kilometres from its source region in Antarctica. By contrast, the water at around 800m depth still shows some of the features of the AAIW, but this water mass has been considerably 'degraded' by mixing with surface water above and with deeper water below.

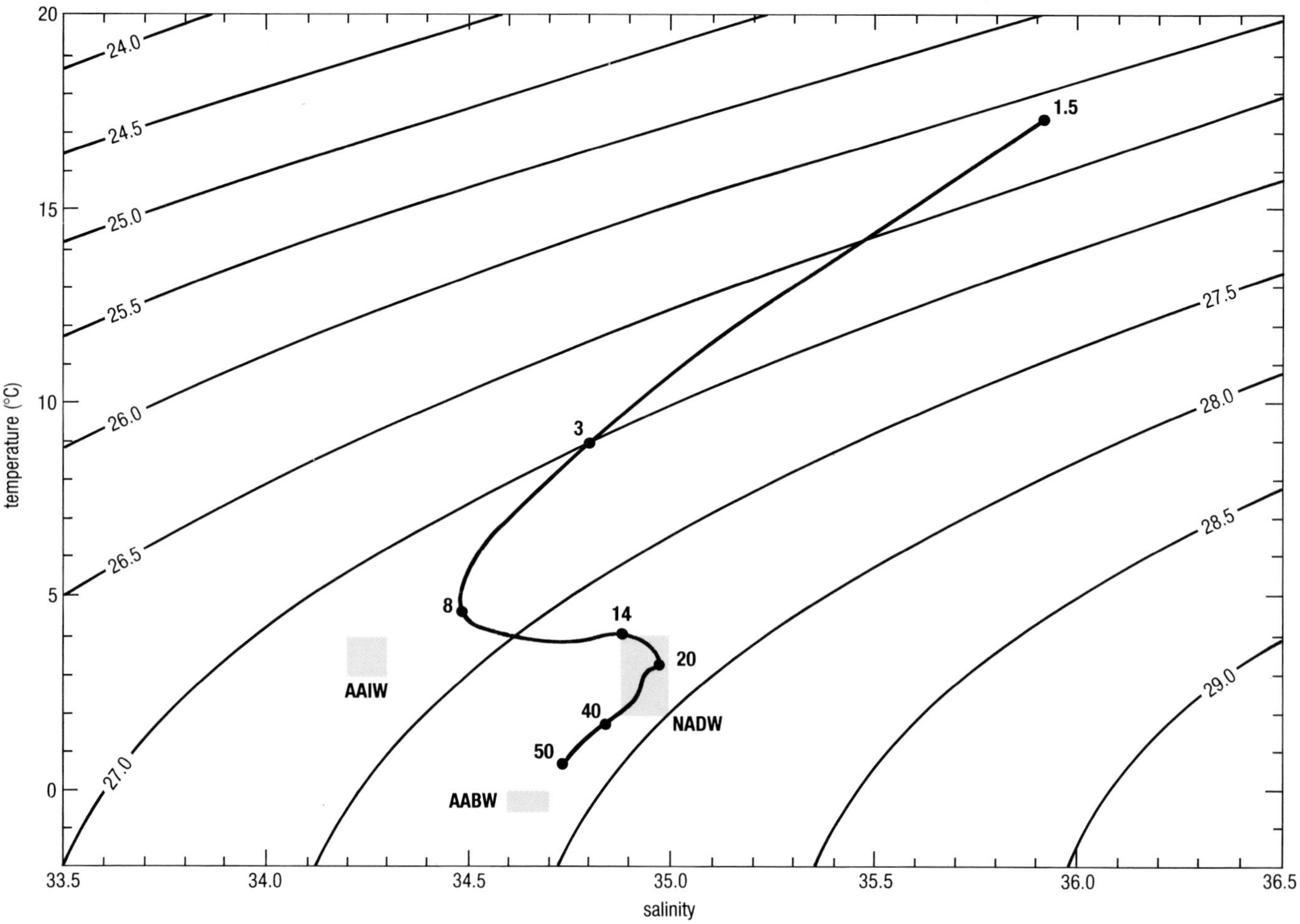

Figure 4.7 An example of a *T–S* diagram for observations from 150 m to 5 000 m depth at a location 9° S in the Atlantic Ocean. Dots represent individual seawater samples; numbers are depths in hundreds of metres. Blue shaded boxes represent the major subsurface Atlantic water masses. AABW=Antarctic Bottom Water; NADW=North Atlantic Deep Water; AAIW=Antarctic Intermediate Water.

QUESTION 4.7 (a) If you were to plot the σ_t values on Figure 4.7 against depth, would the result indicate that the water column is stable?

(b) Why would this result be only a rather rough indication of stability? Would a plot of σ_θ against depth give a more reliable indication?

(c) How would the curve on Figure 4.7 look if potential temperature were used to plot the points instead of *in situ* temperature? Assume that in this case the contours would be for σ_θ values.

4.3.4 CONSERVATIVE AND NON-CONSERVATIVE PROPERTIES

There are two main reasons why the *T–S* diagram is a powerful tool for identifying and tracking water masses. First, temperature and salinity are quite easily measured. Secondly, as soon as the water is out of contact with the atmosphere, i.e. once it has left the mixed surface layer and is in the main body of the ocean, *these properties can only be changed by mixing with water of different T and S characteristics*. For this reason, *T* and *S* are known as **conservative properties**.

Bearing in mind the definition you have just read, would you say that, strictly speaking, potential temperature, θ, is a true conservative property, whereas *in situ* temperature is not?

Strictly speaking, the answer is yes. *In situ* temperature can be changed by processes other than mixing, namely by adiabatic compression or expansion. Potential temperature has been corrected for this effect, so it is the true conservative property. For this reason, the use of *T–S* diagrams is being increasingly replaced by the use of θ–S diagrams.

Water masses can be identified also by chemical and biological characteristics, for example by their content of dissolved oxygen or nutrient salts; and, in the case of upper water masses in particular (*cf.* Figure 4.1), by the presence of certain communities of organisms. Clearly, however, all of these properties can be changed by processes other than mixing, especially biological processes, and so they are called **non-conservative properties**.

It is crucial to remember that these definitions apply only *within the main body of the oceans*. At the boundaries with atmosphere and sea-bed, there are gains or losses of heat, salt or freshwater, by solar radiation, rainfall, river inflow, crustal heat flux, and so on. The distinction between conservative and non-conservative properties and behaviour is extremely important in oceanography and you will encounter applications elsewhere in this Series.

Hydrothermal circulation associated with sea-floor spreading and submarine volcanism supplies large volumes of heated water into oceanic bottom waters, especially at spreading axes. Does this process invalidate the definition of conservative properties given above?

Not at all. The water expelled from hydrothermal vents has very different values of temperature and salinity from those of surrounding bottom waters. Temperature and salinity are conservative properties, and so they can be used to track the subsequent movements of hydrothermal waters in just the same way as is done for the major water masses.

QUESTION 4.8 (a) Look back to Section 3.1 and recall the discussion of departures from constancy of composition for some major constituents. Do those constituents behave conservatively or non-conservatively in seawater?

(b) Is chlorinity a conservative property?

4.4 MIXING PROCESSES IN THE OCEANS

Inhomogeneities in the oceans can occur on a variety of scales, the largest being that of the water masses mentioned earlier in this Chapter. We look at smaller-scale inhomogeneities later in this Section. Mixing processes act to even out the inhomogeneities: they encompass the extremely slow process of molecular diffusion and the much more rapid process of turbulent mixing.

4.4.1 MOLECULAR DIFFUSION AND TURBULENT MIXING

Even in a fluid that is absolutely at rest, if a dissolved substance is unevenly distributed within it, the substance will diffuse down the concentration gradients to even out the distribution. This is **molecular diffusion**, resulting from the motion of the individual molecules. An even distribution of heat is achieved in a similar way: in regions of higher temperature the molecules have higher kinetic energies. Molecular

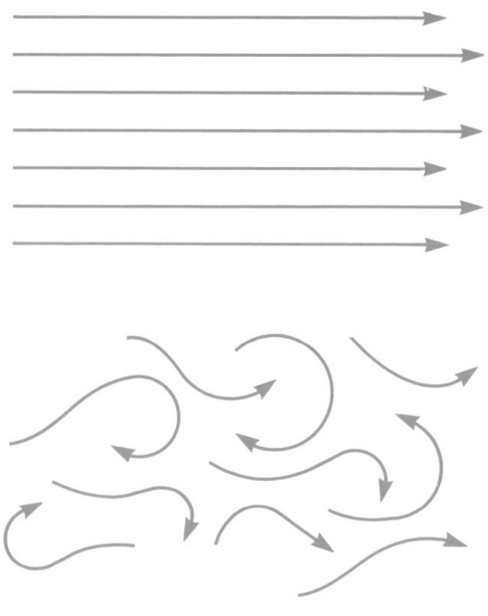

Figure 4.8 Diagrammatic illustration of the difference between (a) laminar flow and (b) turbulent flow.

diffusion of heat occurs as these higher-energy molecules move (diffuse) into regions of lower temperature where they 'mix' with slower-moving molecules and transmit some of their excess energy to them. This is how the process of conduction operates *in a fluid*.

Water in the oceans is usually moving, rarely in laminar flow, most commonly in turbulent fashion. The distinction between the two is shown in Figure 4.8. When fluid is moving by laminar flow, mixing occurs mainly by molecular diffusion. Turbulence (Figure 4.8(b)) can bring waters with very different characteristics into close proximity. **Turbulent mixing** is bulk mixing, like sloshing the water about in a bath, which very quickly achieves a uniform temperature and an even distribution of bath salts.

In the oceans, turbulence may be associated with a wide range of processes: wave motion; vertical or lateral **current shear** (i.e. variations of velocity either with depth or across the flow); water movement over an irregular sea-bed or along an irregular coast; tidal currents, which vary with time as well as with position; and travelling eddies associated with currents.

The oceans are much broader than they are deep—up to about 10000km across, compared with about 5km deep—and horizontal gradients of temperature are many orders of magnitude less than the corresponding vertical gradients. Temperature can change by 10°C or more in 1km depth, whereas it is commonly necessary to travel thousands of kilometres horizontally to experience a temperature change of 10°C. The scale of horizontal turbulent mixing is thus greater than that of vertical turbulent mixing, which in any case tends to be opposed by the vertical stability that results from the increase of density with depth. In short, the effect of density stratification is to inhibit or suppress vertical mixing.

4.4.2 STRATIFICATION, MICROSTRUCTURE AND MIXING

Instruments that can provide continuous profiles of temperature and salinity in the oceans reveal fine-scale features that are known as **oceanic microstructure**. Step-like profiles, in which homogeneous layers of water are separated by thin interfaces with steep gradients of temperature and salinity, have been found in many regions (Figure 4.9). The scale of these features varies considerably, some layers being 20–30m thick (Figure 4.9(a)), whereas others, perhaps superimposed on them, are only 0.2–0.3m thick (Figure 4.9 (c) and (d)). The lateral extent of the layers may be as much as tens of kilometres for the thicker layers and perhaps hundreds of metres for the thinner layers. Temperatures may either increase or decrease with depth in these step-like profiles, but where the temperature increases with depth (a temperature inversion) salinity also increases with depth, otherwise the interfaces between the layers would not remain stable. When the temperature decreases with depth, salinity may either increase or decrease with depth.

Because the density increases across each step, the microstructure is vertically stable, and this tends to inhibit vertical mixing. However, molecular diffusion alone would eliminate differences between adjacent layers of water, given sufficient time. Such diffusion could occur by a mechanism known as **salt fingering**.

Molecular diffusion of heat is many times more rapid than that of salt. If, therefore, we initially have a two-layer system, where less dense warm

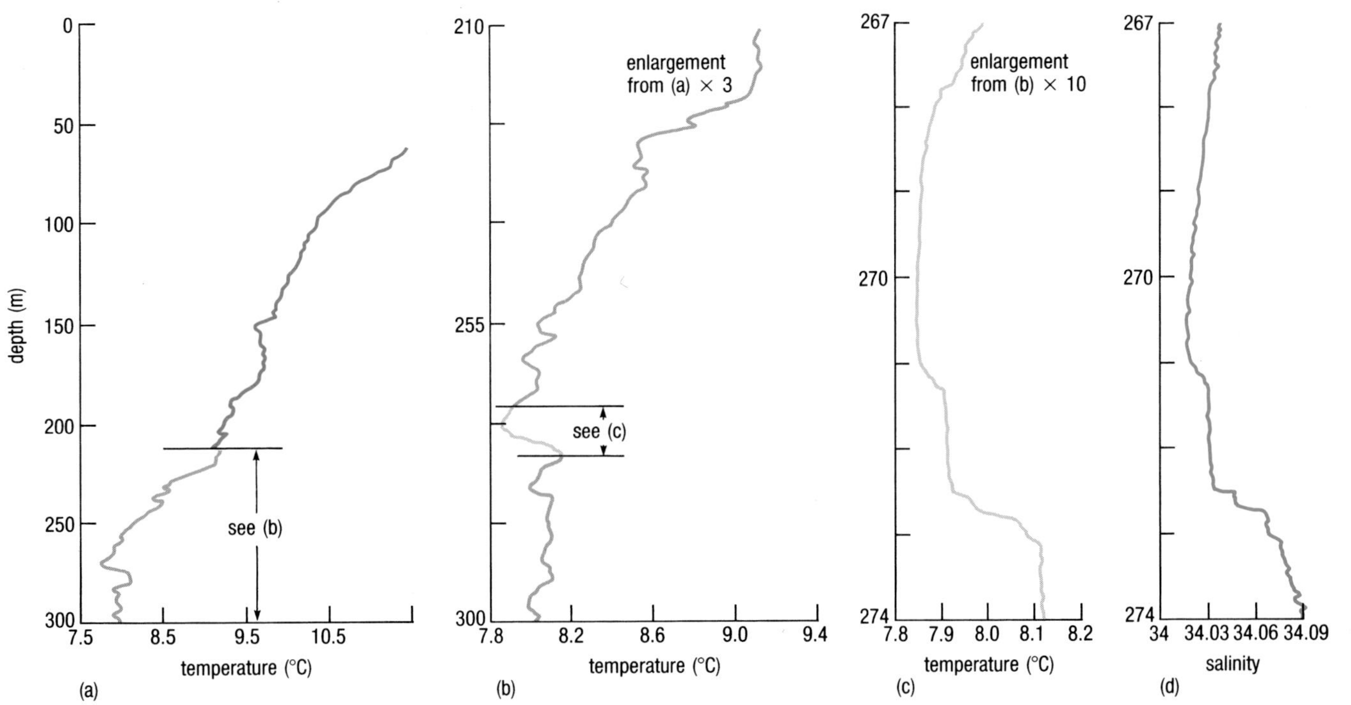

Figure 4.9 Examples of step-like profiles of temperature — (a), (b), (c) — and salinity (d), from a location off the coast of California. Profiles (a)–(c) are successively expanded, to show the fine scale of stratification that can be detected. Microstructure can occur at any depth, but is most common within and above the main thermocline.

salty water overlies denser, cooler and less salty water, the heat diffuses downwards more rapidly than salt. Figure 4.10 shows how this process reduces the density of the lower layer and increases that of the upper layer, leading to instability in the system. The result is a convection pattern of sinking cells of salty water alternating with rising cells of less salty water.

However, the persistence of sharp boundaries between the layers in oceanic microstructure suggests that there is some process which acts to maintain the contrasts across them, counteracting the effects of molecular diffusion. Various hypotheses have been proposed to account for oceanic microstructure and for the processes that act to maintain it. It may well be that different processes dominate on different scales in different parts of the oceans. Here we describe what many oceanographers now believe to be the most likely mechanism for the maintenance of oceanic microstructure.

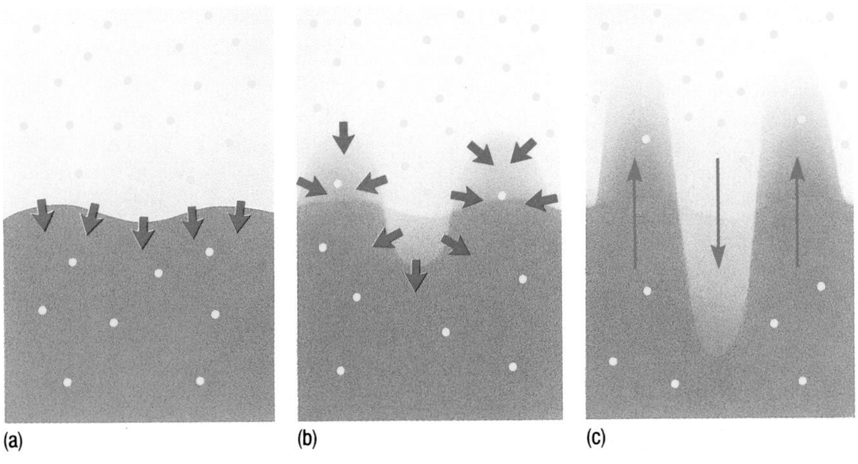

Figure 4.10 The change from a stable to an unstable density–depth profile when warm saline water (light blue) overlies cooler and less saline water (dark blue).

(a) The more rapid diffusion of heat (*short arrows*) than salt, leads to (b) and (c), the development of salt fingers (*long arrows*) when the density profile becomes unstable.

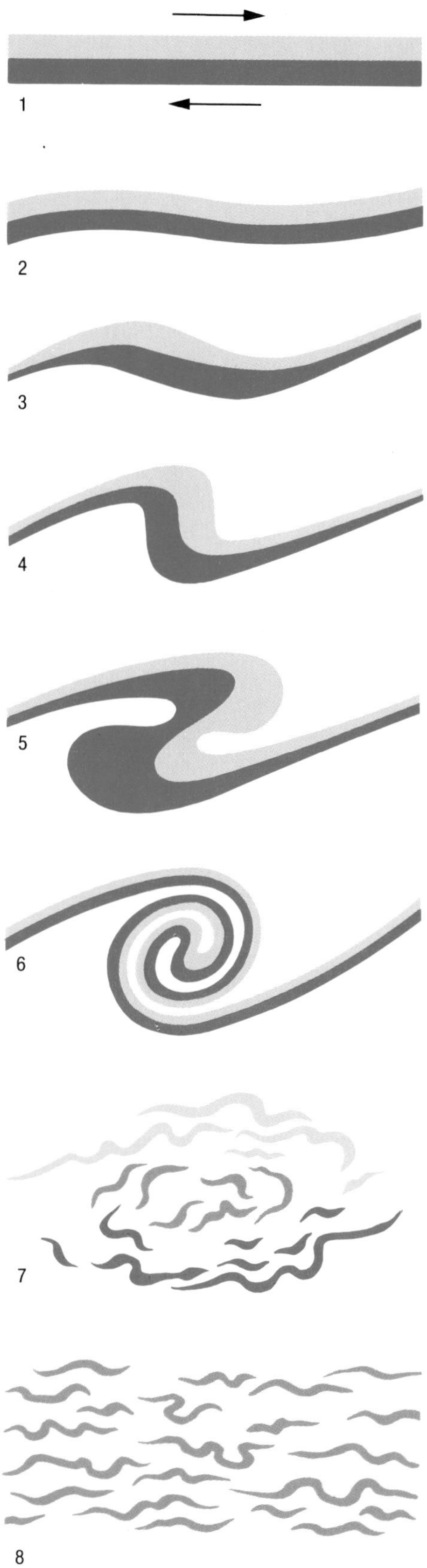

The breaking of internal waves

We have already established that as the density increases downwards across each step in the microstructure it is vertically stable. Wherever the water in the oceans is vertically stable, oscillations can occur if the water is displaced vertically. **Internal waves** result, which can propagate energy through the ocean in the same way as surface waves do.

Such waves can form at the interfaces between layers of different density which are associated with velocity shears, i.e. where the water above and below the interface is either moving in opposite directions or (more likely) in the same direction at different speeds. These shears can produce local instability in the form of billows or breakers (Figure 4.11) which lead to vigorous turbulent mixing of water immediately above and below the interface. The effect of this is to create an intermediate layer between the two original layers, and thus form two smaller steps in the vertical profile in place of one larger step. This can continue indefinitely with further steps in the vertical profile being formed on each occasion.

One of the first observations of this process was in the early 1970s, when the use of dye tracers enabled divers to observe internal waves breaking in the thermocline off Malta. Internal waves in general occur on a variety of scales and are a widespread phenomenon in the oceans. Probably the most important are those associated with tidal oscillations along continental margins. These are large enough to be detected easily on aerial photographs and satellite imagery, provided they are not too deep.

4.4.3 FRONTS

In the oceans, **fronts** are inclined boundaries between different bodies of water having contrasted characteristics. They are analogous to atmospheric fronts between different air masses, and occur on a variety of scales. They can form both within estuaries (between river water and higher salinity estuarine water) and off the mouths of estuaries (between estuarine water and normal seawater). They are common in shallow seas, separating stratified water from water that is vertically mixed; and along continental shelf margins, separating coastal or shelf water from water of the open ocean. On a still larger scale are fronts in the deep ocean between water masses of different properties.

In shelf seas, tidal currents have appreciable velocities close to the sea-bed, and can be an important agent of vertical mixing. If there is a large vertical current shear due to friction at the sea-bed (Figure 4.12), the resulting turbulence leads to the development of a mixed lower layer. If the top of this lower layer merges with the base of the upper mixed layer, the water becomes vertically homogeneous—a common situation in the seas around Britain which are subject to fairly strong tidal currents ($>0.5 \text{ms}^{-1}$). In some areas, however, the tidal currents are weaker, or the total depth of water is greater, and in these areas stratification does develop. Fronts in shelf seas are boundary regions between homogeneous

Figure 4.11 Mixing caused by the passage of an internal wave. Stage 1 shows a layer of lower density overlying and moving faster than one of higher density, so that the *relative* speeds are in opposite directions (arrows). In the succeeding stages (2–8) the two layers lose their coherence as internal waves develop, and break up into turbulent patches. The patches are rapidly flattened by stratification, which gives rise to a finely layered microstructure (*cf.* Figure 4.9) as the movement subsides.

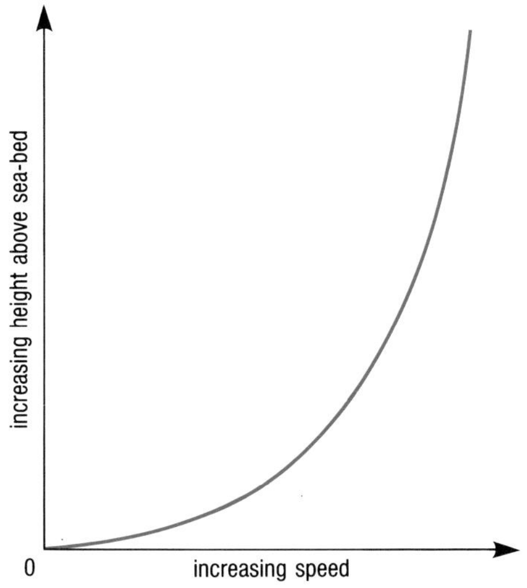

Figure 4.12 Variation in speed with height above the sea-bed to illustrate the principle of vertical current shear; each 'layer' moves faster than the one immediately below it.

(i.e. completely mixed) and stratified waters (Figure 4.13), where the balance between layering and mixing depends mainly upon the strength of the tidal currents.

Why is stratification more likely to develop in mid-latitude shelf-sea areas in summer rather than in winter?

Greater insolation in summer months leads to warmer and less dense surface waters, and mixing is not so intense because in general wind speeds are lower. A seasonal thermocline develops (Section 2.3). In winter, cold weather and generally stronger winds cool the surface layers, which become denser and less stable, and hence more susceptible to mixing by the winds. The thermocline is pushed deeper (*cf.* Figure 2.7(d)) and eventually intersects the top of the lower mixed layer; the whole water column will then be mixed throughout.

The essential feature of a front is the density difference between the water on either side, but other features usually enable it to be seen fairly easily. The front itself is frequently marked by a line of foam or floating debris, because fronts are regions where surface waters converge, i.e. move towards one another on either side of the boundary. The convergence results in part from wind at the surface, but it is also the result of the density contrasts across the front itself.

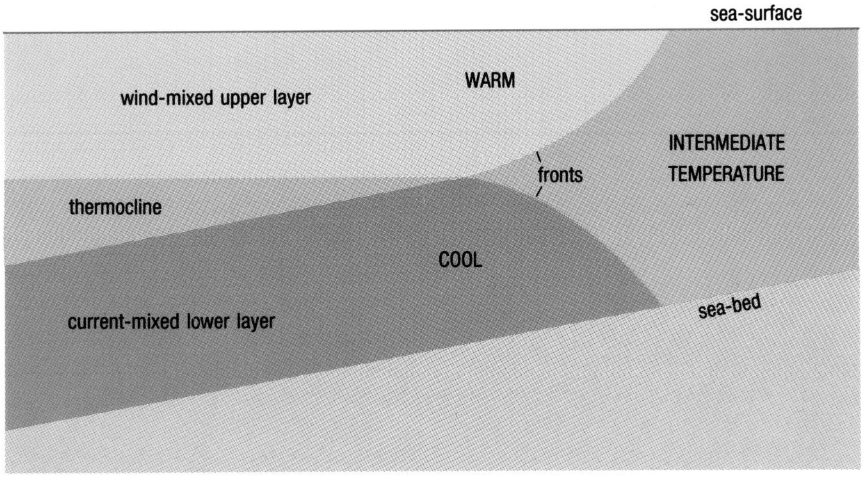

Figure 4.13 An illustration of how fronts can develop between homogeneous waters (right) and stratified waters (left) in shelf seas. (Note that the fronts will in general be gradational zones rather than sharply defined boundaries.) The lower mixed layer is due to tidal currents, the upper mixed layer is due to mixing by wind, and its lower boundary is a seasonal thermocline (probably coinciding with a pycnocline). The vertical scale is greatly exaggerated.

Figure 4.14 illustrates in a simplified way how the density contrasts lead to convergence. Fronts by definition separate water of different density along inclined boundaries. There are strong density gradients *across* them, and so fronts are defined by closely spaced (imaginary) surfaces of equal density—**isopycnal surfaces**. Because these isopycnal surfaces are themselves inclined, the water simply 'slides' downwards along them. If that seems hard to envisage, think of it like this. An isopycnal surface is one of constant density: all points on that surface represent the same density value. If the isopycnal is horizontal, the water will not move; but

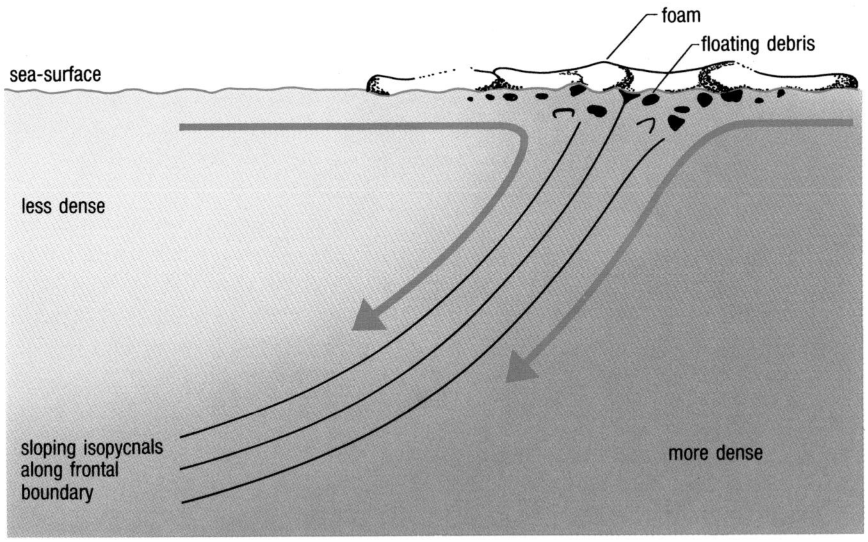

Figure 4.14 Schematic illustration of the convergence of surface water along a frontal boundary. Note the strong density gradients *across* the front which is represented by sloping isopycnal surfaces. The vertical scale is greatly exaggerated. For further explanation, see the text.

if the isopycnal slopes, then water can slide down along it to some depth where it becomes horizontal again. The sinking water draws more water in from above to maintain the supply.

Because the water properties on either side of a front are different, they can be identified easily on aerial photographs and satellite images, especially where there is a change in surface roughness and hence in optical reflectivity. The temperature of the water is nearly always significantly different on each side, and the less-stratified (better-mixed) cooler water on one side is more likely to be richer in nutrients than the more-stratified, warmer water on the other. As a result, fronts can often be recognized by the extent of biological production as well as by temperature distribution, and the two are often well correlated (Figure 4.15).

Fronts can also develop in association with lateral velocity shears, within or near the boundaries of current systems, where adjacent bodies of water are moving in the same direction but at different speeds. They can be very sharp boundaries. For example:

> 'The Gulf Stream, as it issues from the Straits of Florida and expands into the ocean on its northward course, is probably the most glorious natural phenomenon on the face of the Earth. The water is of a clear crystalline transparency and an intense blue, and long after it has passed into the open sea it keeps itself apart, easily distinguished by its warmth, its colour, and its clearness; and with its edges so sharply defined that a ship may have her stem in the clear blue stream while her stern is still in the common water of the ocean.'
>
> C. Wyville-Thompson (1873) *The Depths of the Sea*, Macmillan, p. 382.

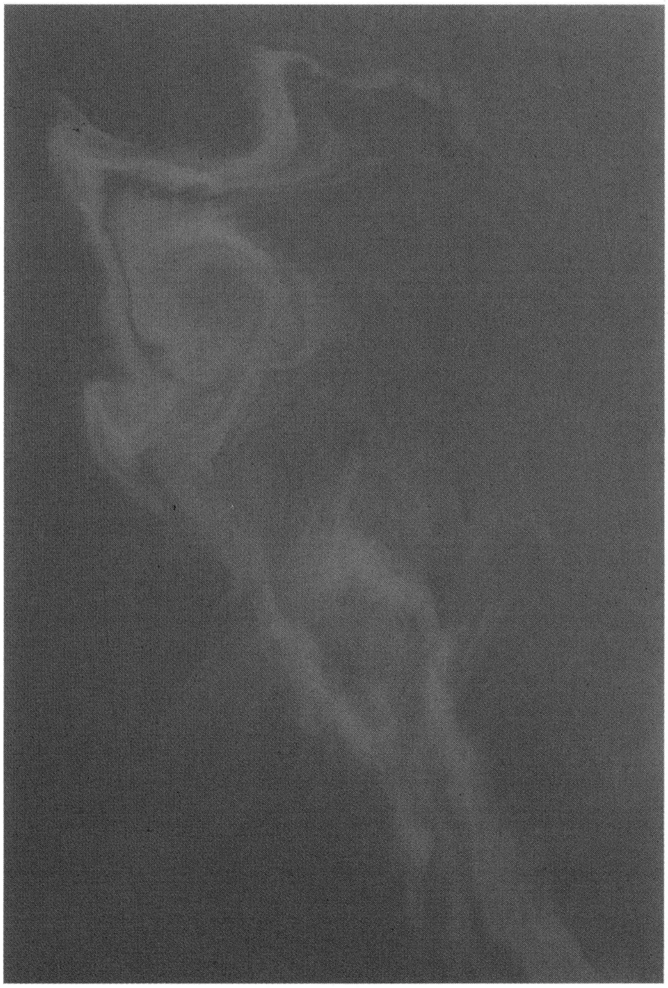

Figure 4.15 (a) A front in the Falklands Current, shown up by contrasted production of plankton. To the right of the front, the paler colours represent high populations of plankton (the result of a phytoplankton bloom), while on the left, the ocean is relatively barren. The front shows the characteristic swirls and eddies found in nearly all frontal regions and at least in part related to the changes of current velocity (lateral velocity shears) across the front. The distance from top to bottom of the photograph is about 100 km.

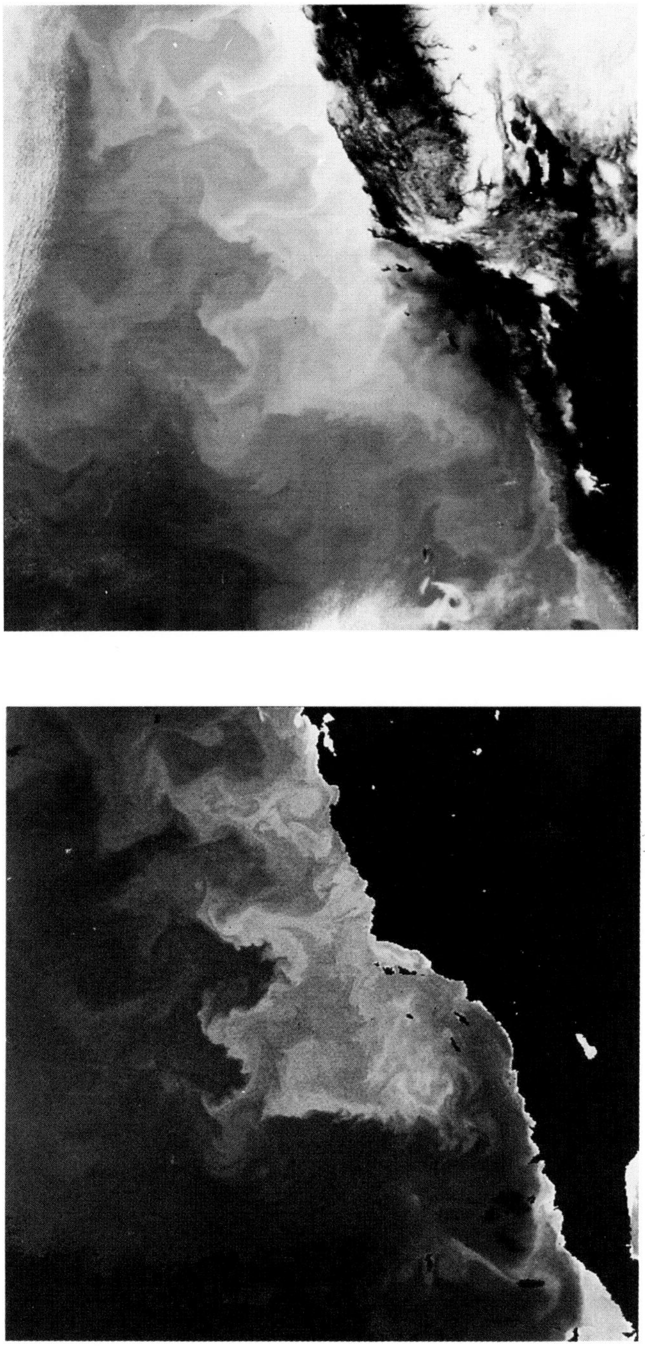

(b) The correlation between sea-surface temperature (*top*) and biological productivity (phytoplankton, *bottom*), on each side of a front off southern California, near the edge of the continental shelf. Here, the waters of the continental shelf are both warmer and more nutrient-rich (possibly because of the proximity of land) than the offshore waters. This picture also shows how swirls and eddies can develop along fronts. The distance from top to bottom of each picture is 700 km.

The boundaries of the Gulf Stream are normally not so abrupt as that: the so-called 'Cold Wall', separating the warm waters of the Gulf Stream from cooler waters on the landward side, is typically a frontal *zone*, some 30–50km wide, over which the temperature changes by about 10°C.

Across most major fronts, however, the temperature gradient is typically much less: of the order of 2°C in about 20km. Smaller fronts in estuarine and coastal waters are sharper (*cf*. Figure 4.14). It is important to stress that *all* fronts slope at very low angles—of the order of 1 in 100—which means that the vertical scales in Figures 4.13 and 4.14 are highly exaggerated.

As you might expect, mixing occurs across fronts, and is an important consideration in, for example, the exchange of coastal and open ocean waters, because the extent of mixing will control (among other things) the removal of pollutants to the deep ocean. Mechanisms of mixing include the development of eddies where there are vertical and/or lateral velocity shears; and the interleaving of the two water masses on either side of the front, producing a 'frontal microstructure' where small-scale mixing processes of the kind described in Section 4.4.2 might operate.

Up to now we have been concerned mainly with what may be called intrinsic properties of seawater. In the next Chapter we look at how these properties affect the propagation of light and sound in the oceans.

4.5 SUMMARY OF CHAPTER 4

1 Water masses are analogous to air masses. They can be identified by characteristic combinations of temperature and salinity and other properties. The boundaries of major upper water masses correspond approximately to the major wind-driven surface current systems. Subsurface water masses have comparatively narrow ranges of temperature and salinity, inherited from surface waters in the source regions where they form and sink by virtue of their increased density. The movement of subsurface water masses is density-driven; this is the thermohaline circulation.

2 Temperature and salinity together control density, but pressure is also an important factor. Pressure increases linearly with depth in the oceans, because water is virtually incompressible. A pressure of about 1 atmosphere (10^5N m^{-2} or 1000mb) is exerted by a 10m water column. Air cools adiabatically as it rises, due to expansion with falling pressure. Water is heated adiabatically as a result of increased pressure and slight compression with depth. The potential temperature (θ) of a water sample is its measured *in situ* temperature after correction for adiabatic heating.

3 Sigma-*t* (σ_t) represents the density of seawater samples at atmospheric pressure, in terms of salinity and *in situ* temperature. Sigma-θ (the potential density, σ_θ) expresses the density of seawater samples in terms of salinity and potential temperature. *T–S* diagrams are countoured in values of σ_t and are used to identify water masses and to determine the extent of mixing between them. Pycnoclines are regions where density increases rapidly with depth and the main pycnocline coincides approximately with the permanent thermocline.

4 Conservative properties of seawater are those which are changed only by mixing, once the water has been removed from contact with the atmosphere and other external influences. Non-conservative properties are those which are changed by processes other than mixing. Temperature (potential temperature) and salinity are conservative properties; dissolved oxygen and nutrient concentrations are non-conservative properties.

5 Mixing occurs by both molecular diffusion and turbulent mixing, the second of which is much the more important in the oceans. Turbulent mixing is much more rapid than molecular diffusion, and horizontal mixing is more important than vertical mixing in the oceans, partly because of their great width:depth ratio, and partly because density stratification inhibits vertical mixing.

6 In many parts of the ocean there is a well-defined microstructure, consisting of layers of water with fairly uniform T and S characteristics, separated by steep gradients of temperature and salinity. A small-scale process that may operate to break down the stratification is salt fingering, which involves the downward molecular diffusion of heat and the upward molecular diffusion of salt. The microstructure is gravitationally stable and is probably both maintained and modified mainly by the breaking of internal waves, the result of velocity shears along density interfaces.

7 Fronts are gently inclined boundaries which separate water of contrasted characteristics, typically well stratified on one side, mixed and more uniform on the other. Major fronts are normally tens of km across and slope down beneath the warmer and more stratified water, often at very small angles. They are common in shallow continental shelf waters, over the continental shelf, along continental margins, and where lateral velocity shears are associated with oceanic current systems. They are characterized by sloping isopycnal surfaces (surfaces of constant density). Near-surface water can sink to greater depths along sloping isopycnals.

Now try the following questions to consolidate your understanding of this Chapter.

QUESTION 4.9 How do processes involved in the formation of ice help to drive the thermohaline circulation of the oceans?

QUESTION 4.10 Temperature generally decreases with depth in the oceans. In the troposphere (the lower 10–15km of the atmosphere), which contains three-quarters of the mass of the atmosphere, temperature decreases with height. Are the causes of the temperature decrease the same?

QUESTION 4.11 From Figure 4.4, for water at 16°C and with a salinity of 34:

(a) What change in *temperature* would cause a change in specific gravity from 1.025 to 1.026 at constant salinity?

(b) What change in salinity would cause the same change at constant temperature?

(c) If a sample of seawater with T–S characteristics of 2°C and 34.5 were collected from a depth of 4 000m, would its σ_t value correspond to a density greater or less than its true density?

QUESTION 4.12 Examine Figure 4.16. Large areas representing T and S values for the three main oceans (Pacific on the left, Atlantic on the right, Indian in the middle) converge near the bottom of the diagram.

(a) Does the large spread of values of T and S for each ocean represent mainly near-surface or mainly deep waters?

(b) Table 4.2 presents estimates of *average* temperature and salinity values for each of the main ocean basins, and the average for the world's oceans as a whole. Plot on Figure 4.16 the values for each ocean. Can you suggest why the three points cluster together where they do?

Table 4.2 Average temperatures and salinities for the major ocean basins.

Ocean	Temperature (°C)	Salinity
Pacific	3.36	34.62
Atlantic	3.73	34.90
Indian	3.72	34.76
all oceans	3.52	34.72

Figure 4.16 A T–S diagram for waters of the world's major ocean basins. (For use with Question 4.12.)

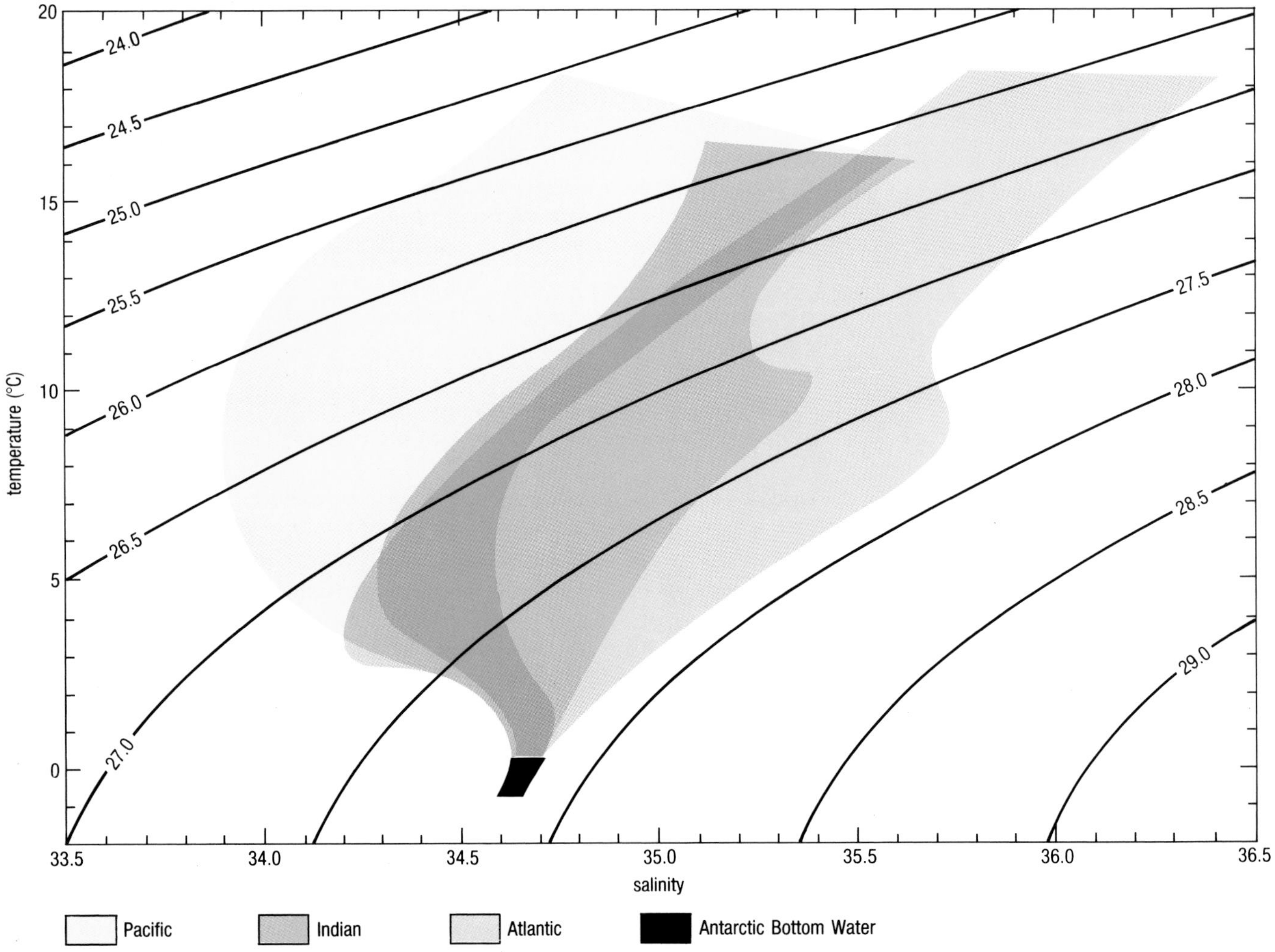

QUESTION 4.13 Are each of statements (a)–(d) true or false?

(a) Salt fingering is a more important mixing process than internal wave breaking.

(b) According to the definition, *in situ* temperature is not a conservative property, but potential temperature is.

(c) When it is not possible to measure the *in situ* temperature, the potential temperature is measured instead.

(d) Both surface heating of the oceans and precipitation (rainfall) promote stability of the surface layers.

QUESTION 4.14 This question concerns the formation of sea-ice and draws on material from both Chapter 1 and Chapter 4. As surface seawater cools, its density increases and it sinks (*cf*. Figure 1.2). In seawater of normal salinity, sea-ice can therefore only begin to form when the whole water column is cooled to about −1.9°C. This is not likely to happen in ocean basins that are several km deep. At first sight, then, it would seem that sea-ice could only form in shallow water; but in the Arctic and Antarctic Oceans it can form offshore where low salinity water overlies water of higher salinity.

(a) By extrapolating from Figure 4.4, make an estimate of the density of water at a temperature of −1°C and with a salinity of (i) 33 and (ii) 35.

(b) Could the water of salinity 33 cool to freezing point without sinking below the higher salinity water beneath it?

CHAPTER 5 LIGHT AND SOUND IN SEAWATER

Humans are generally accustomed to consider sight to be a more important sense than hearing. Light travels faster and penetrates further through the atmosphere than does sound, so we can make better use of our sight, and of electromagnetic radiation generally, in making scientific observations. For animals in the oceans, by contrast, hearing is the more important sense. Sound travels well through water, and this makes possible the remote sensing of objects (e.g. echo-sounding) and the transmission of information (e.g. the 'singing' of whales). The speed of sound in water remains less than that of light, but light can travel only relatively short distances through water. Therefore, the greater part of the oceans is almost completely dark.

5.1 UNDERWATER LIGHT

Light is a form of electromagnetic radiation, which travels at a speed close to 3×10^8 m s^{-1} in a vacuum (reduced to about 2.2×10^8 m s^{-1} in seawater). Oceanographers are interested in underwater light in two main contexts: vision and photosynthesis.

When light is propagated through water, its intensity decreases exponentially with distance from the source. The exponential loss of intensity is called **attenuation** and it has two main causes:

1 **Absorption:** This involves the conversion of electromagnetic energy into other forms, usually heat or chemical energy (e.g. photosynthesis). The absorbers in seawater are:

(a) Algae (phytoplankton) using light as the energy source for photosynthesis.

(b) Inorganic and organic particulate matter in suspension (other than algae).

(c) Dissolved organic compounds (see Section 5.1.4).

(d) Water itself.

Note: (a) and (b) are collectively termed the **seston** (*cf.* Section 6.1).

2 **Scattering:** This simply changes the direction of the electromagnetic energy, as a result of multiple reflections from suspended particles. Scattering by all but the very smallest particles is generally forwards at low angles—i.e. the paths of most of the scattered light are deflected only slightly from the original direction of propagation. Obviously, the greater the amount of suspended matter (i.e. the more turbid the water) the greater the degree of absorption and scattering.

Coastal waters tend to be particularly turbid. The suspended load brought in by rivers is kept in suspension by waves and tidal currents, which also stir up sediment already deposited on the bottom. In addition, rivers supply coastal waters with nutrients that support phytoplankton growth, and with dissolved organic compounds (item (c) in the list of absorbers above). By contrast, the water tends to be particularly clear in central oceanic regions, especially where concentrations of nutrients are low and there is little biological production.

QUESTION 5.1 (a) According to Figure 5.1, is the intensity of light sufficient for phytoplankton to grow (i) on a moonlit night; (ii) at depths greater than 100m in sunlit coastal waters; (iii) at depths less than 200m in clear sunlit ocean waters?

(b) According to Figure 5.1, could fish living in the oceans at a depth of 1000m perceive: (i) moonlight; (ii) sunlight?

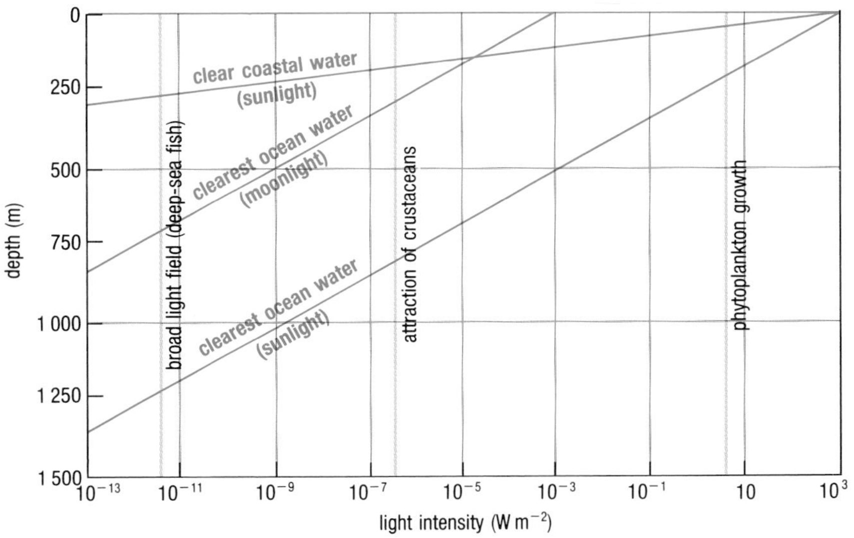

Figure 5.1 The relationship between illumination and depth. Pale blue lines show the light intensity required for various functions. The 'broad light field' for deep-sea fish indicates the minimum quantity of general daylight that these fish can perceive; the intersection with the 'clearest ocean water (sunlight)' line at a point corresponding to about 1 250 m indicates that below this depth fish cannot perceive daylight. More light is needed to attract crustaceans, and still more for phytoplankton growth. (For comparison, the lowest intensities that the human eye can perceive are of the order of 10^{-12} W m^{-2} for a small light source and 10^{-8} to 10^{-9} W m^{-2} for a broad, diffuse light source.)

The illuminated zone in which light intensities are sufficient for photosynthetic primary production to lead to net growth of phytoplankton is called the **photic zone** (or euphotic zone). The greater the clarity of the water and the higher the Sun is in the sky, the greater the depth to which light penetrates and the greater the depth at which photosynthesis can proceed. The photic zone can therefore be up to 200m deep in clear waters of the open ocean, decreasing to about 40m over continental shelves, and to as little as 15m in some coastal waters. Only when the sea-bottom is shallow enough to be included in the photic zone are bottom-dwelling or benthic plants (e.g. attached seaweeds) able to grow—elsewhere in the ocean all plant life must float, i.e. it is planktonic. The wavelength of light is also important in photosynthesis (see Section 5.1.4).

Between the photic zone and the ocean floor is the **aphotic zone** where plants cannot survive for long, because light intensities are not sufficient to enable photosynthetic production to meet the requirements of respiration (see also Section 6.1.3). Below about 1000m depth in the oceans, daylight can no longer be perceived (Figure 5.1, Question 5.1 (b)). This means that throughout most of the oceans there is no external light at all. The only light is that provided by those fishes and other animals that possess bioluminescent light organs (and by human explorers using submersibles and other equipment).

5.1.1 ILLUMINATION AND VISION

In the photic zone and upper parts of the aphotic zone, objects in the sea are illuminated by sunlight (or moonlight), the intensity of which decreases exponentially with depth from the surface, because it is attenuated by absorption and scattering. This so-called **downwelling irradiance** is diffuse, i.e. non-directional, because light may be scattered away from the object, but can also be scattered towards it. In other words, the light illuminating the object has not all taken the shortest path to it from the surface (Figure 5.2(a)). For an object to be *seen*, however, light emanating from the object must be directional, because a coherent image can only be formed if light travels directly from the object to the eye or camera (Figure 5.2(b)).

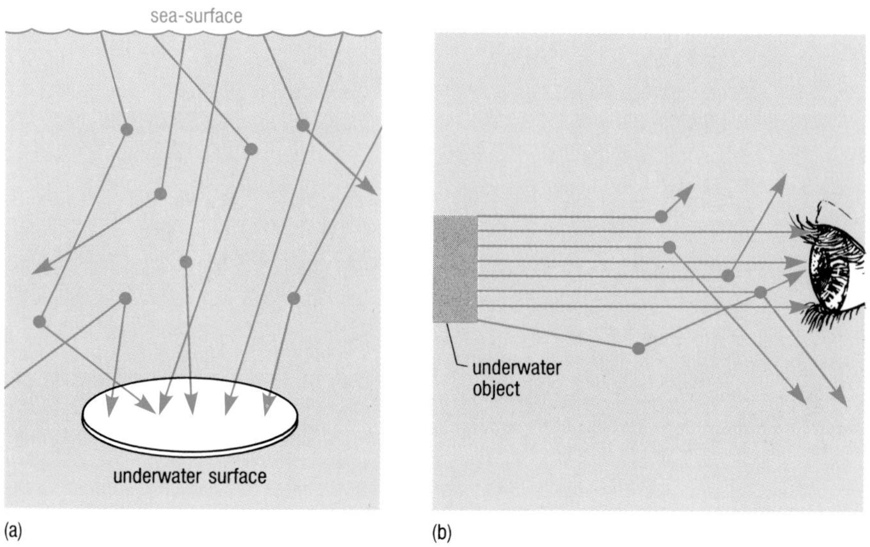

Figure 5.2 Diagrams showing the distinction between (a) the non-directional nature of illumination of an underwater surface and (b) the directional requirements of underwater vision — light *scattered* towards the eye will form part of a coherent image.

The distinction between illumination and vision is nicely illustrated in Figure 5.3. The fish is illuminated by non-directional light, but the image must be transmitted to the diver's eye by directional light for the fish to be seen. Perhaps a more everyday example is provided by a foggy day: your surroundings disappear but your view does not go black—in other words, you have illumination but no vision.

Looking at Figures 5.2 and 5.3, which do you think will be subject to the greater degree of attenuation: the non-directional light that provides illumination, or the directional light required to produce a coherent image in the eye or camera?

Light scattered away from an object being illuminated by the downwelling irradiance is 'compensated for' by light scattered towards the object. Light scattered out of the direct path from object to eye cannot be similarly 'compensated for' because light scattered towards the eye will not contribute to a coherent image, even though it originated from the object. So, it is the directional light associated with vision that is subject to greater attenuation.

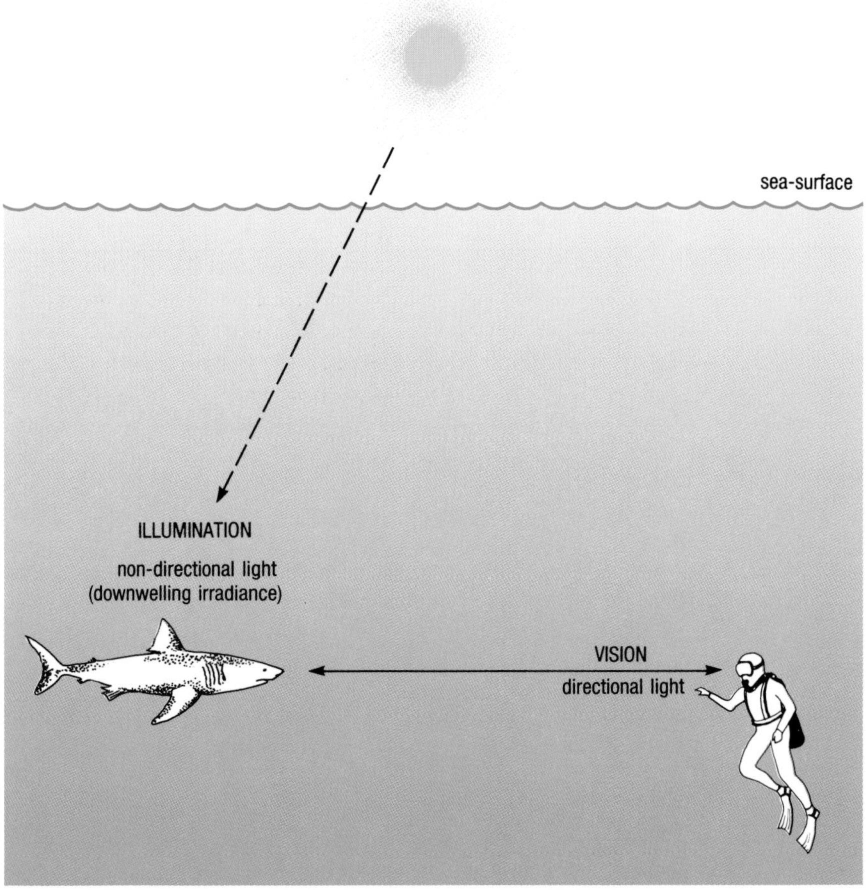

Figure 5.3 Illumination and vision under water. The more turbid the water, the greater the attenuation of light by absorption and scattering, the less the illumination at a particular depth, and the shorter the distance over which objects can be seen.

5.1.2 UNDERWATER VISIBILITY

Visibility is a matter of contrast. An object may show up against its background either because it is a different colour, or because it has a different brightness (or both). Brightness contrast is more important than colour contrast in the marine environment, except in the upper few tens of metres of the photic zone (e.g. in the clear waters of the tropical reef environment, colour contrasts are very important for inter- and intra-specific recognition, camouflage, deterring predators, and so on). At depths greater than a few tens of metres, the downwelling irradiance has not only been much attenuated by absorption and scattering; it has also become almost monochromatic, because of selective absorption of different wavelengths. Accordingly, at the low levels of light typical of most of the underwater world, even the eyes of animals that can normally distinguish colours must use the more sensitive night vision cells, with which everything is seen in shades of grey.

Contrast will decrease with distance, for two reasons: first, the light from the object being observed is attenuated by absorption and scattering; secondly, some of the incoming sunlight (or moonlight) is scattered towards the observer throughout the entire length of the path of sight. This effectively produces a 'veil of light', behind which the object becomes progressively more indistinct, until it disappears against its background.

Figure 5.4 Examples of fishes with luminous organs.
(a) *Vinciguerra attenuata.* (b) *Cyclothone microdon.* (c) *Argyropelecus gigas.*
(d) *Myctophum punctatum.* (e) *Lampanyctus elongatus.* (f) *Bathophilus longipinnis.*
(g) *Argyropelecus affinis.* (h) *Eurypharynx pelecanoides.* (i) adult female with (j) parasitic male of deep-sea angler fish *Ceratias holboelli.*

QUESTION 5.2 Why do you think many fishes living in upper parts of the aphotic zone have dark upper surfaces but silvery undersides?

The light field becomes virtually symmetrical at depths of more than about 250m, which means that the intensity of illumination is similar whether you look upwards or downwards. In the 250–750m depth range, many fishes have silvered flanks, produced by interference 'mirrors' formed out of crystals of guanine (a nitrogenous compound), precisely orientated so as to function vertically when the fish is in its normal upright position. Light is reflected from these 'mirrors' with the same intensity as that of the background, thus effectively presenting zero contrast. Such fishes also have ventral photophores (luminous organs) which break up their silhouettes when viewed vertically from below; while their dorsal regions are black to minimize the contrast when viewed from vertically above—the hatchet fish (*Argyropelecus*) is a common example.

In the upper parts of the aphotic zone—down to about 1000m—where visual contact is still possible, many fish have developed large eyes to cope with the low light intensities. At greater depths, luminous organs arranged in distinctive patterns are developed in species that still depend on sight for contact, and fishes have become a uniform non-reflective black, so that they are not illuminated by the light of others.

In this environment, light is used in all the ways that colour is used in the terrestrial environment, e.g. (see also Figure 5.4):

- to deter would-be predators by appearing larger, e.g. with the aid of lights at the ends of long spines;
- to identify one's own species and/or mate;
- to provide signals whereby shoals can keep together;
- to break up the outline when viewed from below (*cf*. hatchet fish, also many other fish species, as well as squid and euphausid crustaceans); and
- as lures to attract prey (as well as headlamps to illuminate it).

In the lowest parts of the aphotic zone (below about 4000m), luminous organs are less common and eyes correspondingly reduced in size or even absent.

5.1.3 MEASUREMENT

Instruments used for the measurement of underwater light fall into three main categories:

1 Beam transmissometers and turbidity meters measure the attenuation of a parallel (collimated) light beam from a source of known intensity, over a fixed distance. The ratio of light intensity at source and receiver (which are separated by a known distance) provides a direct measure of the attenuation coefficient for directional light.

2 Irradiance meters accept light coming from any direction. The light is usually received by a teflon sphere or hemisphere, which measures ambient light downwelling from the surface—the downwelling irradiance. By making measurements of light intensity at different depths, the attenuation coefficient (called in this case the diffuse attenuation coefficient) for the non-directional downwelling irradiance can be determined.

QUESTION 5.3 Bearing in mind the earlier discussion of Figures 5.2 and 5.3:

(a) Would you expect the attenuation coefficient for directional light to be greater or less than the diffuse attenuation coefficient for non-directional light?

(b) For which of the types of water represented in Figure 5.1 would these coefficients be greatest, and for which would they be least?

As you might perhaps expect, increased turbidity has a proportionately greater effect on directional than non-directional light. The value of the ratio:

$$\frac{\text{attenuation coefficient (directional light)}}{\text{diffuse attenuation coefficient (non-directional light)}}$$

can be less than 3 in the open oceans, but as high as 10 or more in a turbid estuary.

3 Nephelometers provide a direct measure of scattering in the water. A collimated beam illuminates a pre-determined volume of water which scatters light in all directions. The receiver is aimed at the centre of this scattering volume and can be rotated round it, so that variations in the scattering loss with direction relative to the light beam can be determined (Figure 5.5). As the degree of scattering is related to the amount of suspended material in the water, nephelometers can be used to provide a quantitative measure of turbidity, i.e. the concentration of suspended material. Nephelometry has been used for example to determine concentrations of suspended sediment in the deep ocean, and thus to provide information about the distribution and speed of bottom currents.

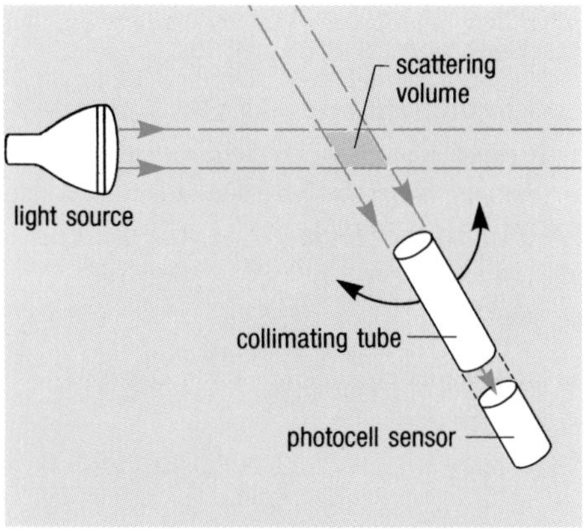

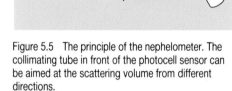

Figure 5.5 The principle of the nephelometer. The collimating tube in front of the photocell sensor can be aimed at the scattering volume from different directions.

The **Secchi disc** is a much more homely piece of equipment. It is simply a flat circular plate, 20–30cm in diameter, either all white or with two quadrants painted black and two painted white (Figure 5.6). It is lowered through the water column in a horizontal attitude until it is observed just to disappear. The depth at which this happens is called the **Secchi depth**, and it depends on the turbidity of the water. The Secchi disc is both cheap and easily made, and it has been used by oceanographers for over a century as a rapid means of assessing water clarity.

Figure 5.6 The Secchi disc.

Simple empirical equations enable a good deal of information to be gleaned from the Secchi depth. The basic relationship for the vertically observed Secchi disc is:

$$Z_s = \frac{F}{C+K} \tag{5.1}$$

where:

Z_s is the Secchi depth.
C is the attenuation coefficient for directional light.
K is the diffuse attenuation coefficient for non-directional light (sometimes also called the extinction coefficient).
F is a factor that depends on the reflectivity of the disc and that of the background, and the observer's own threshold perception of contrast. It is about 8.7 in clear oceanic water, but can be as little as 6 in turbid estuarine water.

The Secchi disc can be used to estimate the diffuse attenuation coefficient, K, of the water. This is the appropriate coefficient for studies of photosynthetic primary production, because it relates to the exponential decrease in intensity of the downwelling irradiance, and hence to the depth of the photic zone. In brief, an empirical relationship between K and Z_s is assumed, such that $K \times Z_s$ = constant. Many workers have measured K and Z_s in different types of water and most have found that $1.4 < K \times Z_s < 1.7$. We can therefore estimate K from Z_s by using:

$$K \times Z_s = 1.5 \tag{5.2}$$

The Secchi disc can also be used to estimate underwater visibility, which relates to the attenuation coefficient, C, for directional light (equation 5.1). Observations in Plymouth Sound have demonstrated a strong empirical relationship between Secchi depth and horizontal underwater visibility (the range at which the contrast of a black object becomes zero and it disappears against its background):

$$V = 0.7\ Z_s \tag{5.3}$$

Those same observations have led to two other empirical relationships that enable the depth of the photic (or euphotic) zone, Z_e, to be estimated:

$$Z_e = 3\,Z_s \tag{5.4}$$

and, because $Z_s = \dfrac{V}{0.7}$

$$Z_e = \frac{3V}{0.7}$$

$$Z_e = 4.3V \tag{5.5}$$

For other regions, the numerical factors in equations 5.3 to 5.5 may well be different.

QUESTION 5.4 (a) If the Secchi depth is 10m, what is the value of the diffuse attenuation coefficient K? What is the potential percentage error on this value?

(b) If Z_s is 20m, what is (i) the horizontal visibility and (ii) the depth of the (eu)photic zone?

Underwater contrast, and hence visibility, depends also on the sighting angle: horizontal visibility is not necessarily the same as visibility looking upwards or downwards. A more general expression is:

$$V = \frac{F}{C + K\sin\theta} \tag{5.6}$$

where V, C, K and F are as before, except that F has a greater range of values than in equation 5.1, because it varies according to the objects being observed, and θ is the sighting angle from the horizontal.

QUESTION 5.5 How does equation 5.6 show that horizontal visibility is a function of C, but not of K?

In conclusion, it is worth recording that neither temperature nor salinity of seawater has any appreciable effect on these phenomena: the coefficients C and K of clear seawater are virtually the same as those of pure water.

5.1.4 COLOUR IN THE SEA

Bearing in mind the information contained earlier in Figure 2.5 and related text, can you suggest why many animals in the upper part of the aphotic zone (Section 5.1) have evolved black or red coloration?

Red animals appear red because they reflect red light, and the only light available from the downwelling irradiance in this 'twilight' zone is blue–green (Figure 2.5 shows that longer wavelengths of the visible spectrum have been absorbed at 100m depth). So red animals will appear black (along with those that really are black) and will therefore be inconspicuous—an advantage for predator and prey alike.

The carotenoid pigments that provide the red colour also have maximum absorbence in the wavelengths emitted by most bioluminescent organs (photophores). This means that red fishes will not show up in the 'headlamps' of others, such as *Diaphus*, which use light organs to illuminate their prey (Section 5.1.2). Some fishes, however, have actually developed light organs that produce red light (e.g. *Pachystomas*) and their eyes contain the visual pigment to detect it. They can see without being seen, because the eyes of most other fishes are adapted to register only blue–green wavelengths, and red coloration is no camouflage when red light shines on it.

Attenuation of underwater light results from a combination of absorption and scattering. Scattering of light by particles is largely independent of wavelength, but absorption is not. The principal absorbers in the sea, as listed in Section 5.1, absorb different wavelengths of light in different proportions.

(a) *Algae*: Chlorophyll 'looks' green because it reflects well at the middle of the visible spectrum; it follows that it must absorb strongly at the two ends. Figure 5.7 contains similar information to Figure 2.5, but for a narrower wavelength band. It compares the energy spectra of solar radiation reaching different depths in various types of water. In brief, blue–green light (450–500nm) penetrates furthest in the open ocean, and in fact about 35% of light of this wavelength that is incident on the surface reaches a depth of 10m. In turbid coastal water, on the other hand, yellow–green light (500–550nm) penetrates deepest, but only about 2% of light of these wavelengths that is incident on the surface reaches a depth of 10m.

Many algae contain pigments that absorb light energy at longer wavelengths and transfer it to the cholorophyll system. For this reason,

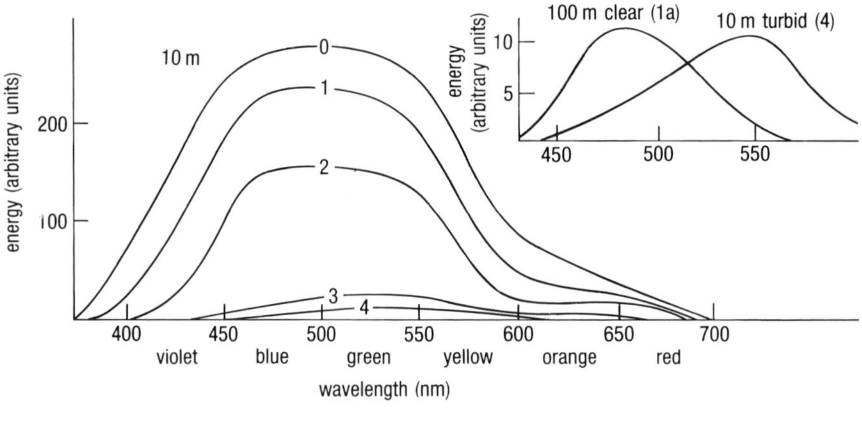

Figure 5.7 Energy spectra at a depth of 10 m for: pure water (0), clear oceanic water (1), average oceanic water (2), average coastal water (3), and turbid coastal water (4).

Inset: An energy spectrum at 100 m depth in clear oceanic water (1a) compared with that for 10 m in turbid coastal water (4).
Compare this Figure with Figures 2.5 and 5.1 and note that it represents only a small part of the spectrum shown in Figure 2.5.

light in the whole wavelength band from 400nm (deep violet) to 700nm (dark red) is described as *photosynthetically active radiation* (PAR). However, most photosynthesizing organisms use blue–green wavelengths (450–500nm), and as we have seen it is these wavelengths that are preferentially transmitted by clear seawater. This correlation is no accident, but is the result of evolutionary selection.

QUESTION 5.6 What does the following statement mean?
If blue–green light is preferred by most photosynthesizing organisms, the selective transmission of yellow light in turbid coastal waters (Figure 5.7) must be of considerable significance to the depth of the photic zone and hence to levels of photosynthetic primary production there.

(b) *Particulate matter*: At normal concentrations, inorganic and other organic particles absorb weakly but scatter strongly. Their comparatively small absorption is mainly in the blue range so their effect tends to be swamped by that of the dissolved organic compounds (see below).

(c) *Dissolved organic compounds*: These are variously known as **yellow substances**, Gelbstoff, or gilvin. During the decomposition of plant tissue, organic material is broken down into CO_2, inorganic compounds of nitrogen, sulphur and phosphorus (the nutrients) and complex humic substances. It is these metabolic products that give some inland waters their distinctive yellow–brown coloration. They are brought to the sea by rivers, but are also produced in oceanic waters by the metabolism of plankton. Yellow substances absorb strongly at the short wavelength (blue) end of the spectrum, and reflect well (low absorption) in the yellow–red, hence the characteristic colour.

(d) *Water*: Water appears as a blue liquid, because absorption at the short wavelength (blue) end of the spectrum is relatively low whilst at the long wavelength (red) end it is high (Figure 2.5). Although water appears colourless in small quantities, its blue colour becomes apparent in clear tropical waters or a clean swimming pool. Absorption is so strong in the red that a 1m-thick layer of pure water will absorb about 35% of incident light of wavelength 680nm.

QUESTION 5.7 What percentage of red light is absorbed by 3m of pure water?

Unproductive oceanic water carries little or no algae or yellow substances. It is therefore ‘pure water blue’ in colour. Blue is sometimes called the ‘desert colour’ of the oceans and is typical of many tropical waters. In recent years, a number of lakes in Scandinavia, Canada, and elsewhere have ‘died’ (allegedly because of acid rain) and become a ‘beautiful tropical blue’. In productive waters, red is absorbed by the water and blue is absorbed by yellow substances. This leaves ‘sea-green’—the typical colour of productive mid-latitude waters.

There is commonly a colour change in the water along frontal boundaries (Section 4.4.3), especially where shelf water is separated from water of the open ocean.

In general, would you expect the spectral shift to be from blue to green or vice versa, in passing from the shelf to the deep water?

Shelf waters normally carry higher concentrations of yellow substances and suspended particles than the deeper waters of the open ocean. So, we might expect the shift to be from green to blue when passing from the shelf to the deep water.

5.1.5 ELECTROMAGNETIC RADIATION AND REMOTE SENSING OF THE OCEANS

Passive remote sensing makes use of naturally reflected visible and near infrared wavelengths, as well as naturally emitted longer wavelength infrared and microwave radiation, to provide information about colour (and hence biological production), temperature and ice cover at the surface of the oceans (e.g. Figures 1.5, 1.6, 2.3, 4.15). It also provides information about surface roughness due to winds, waves, tides and currents, as well as about cloud type and extent, and the amount of water vapour in the atmosphere.

Active remote sensing involves the transmission of microwave pulses (radar) from aircraft or satellites, at wavelengths typically of a few cm, followed by measurement and analysis of the signals reflected by the surface. Imaging radar techniques provide information about sea-surface roughness (wave patterns and wave distribution) and ice cover. Radar has the advantage that it penetrates clouds and is capable of providing high resolution.

We have seen that electromagnetic radiation can travel only short distances through water, so remote sensing and aerial photography provide direct information only about the ocean surface (though wave and ripple patterns can vary according to bathymetry, which can therefore sometimes be inferred from radar images). It follows also that communication by radio is not possible under water, even though the attenuation coefficient for longer wavelength radio waves is less than it is for light. It is in fact possible to communicate with submarines submerged at depths of not more than a few tens of metres, either by using very long wavelength (very low frequency, VLF) radio waves, or by using laser beams from satellites. Laser light is very intense and in the 450–500nm wavelength band it can penetrate far enough below the surface to be useful, before its energy is lost by attenuation. But that is the limit to which electromagnetic radiation can be used in the oceans. For both remote sensing and communication *within* the oceans, therefore, it is necessary to make use of much slower-moving acoustic radiation.

5.2 UNDERWATER SOUND

Although both light and sound can be considered to travel as waves, they are fundamentally different. As stated in Section 5.1, light is a form of electromagnetic energy. It propagates most effectively through a vacuum and in general less well as the density of material increases. Sound or acoustic energy involves the vibration of the actual material through which it passes and thus, in general, propagates best through solids and liquids, less well in gases and not at all in a vacuum.

In short, sound is a form of pressure wave, propagated by vibrations that produce alternating zones of compression (molecules closer together) and rarefaction (molecules further apart) (Figure 5.8(a)). All sounds result from vibrations (e.g. the vibrating membrane of a loudspeaker or the

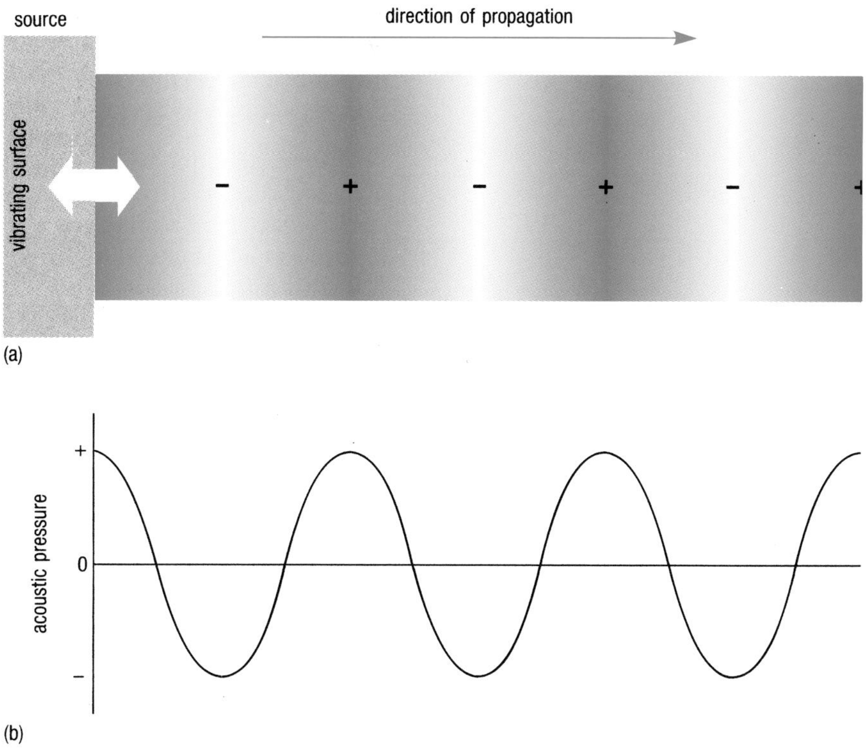

Figure 5.8 Characteristics of the acoustic wave. (a) Propagation of alternating zones of compression and rarefaction. (b) Sinusoidal rise and fall of acoustic pressure as the sound wave passes.

vibrating sound organ of a deep sea animal). Sound waves are thus not sinusoidal in the way that we normally consider wave motions to be. However, the *acoustic pressure* rises and falls in a sinusoidal manner as the sound wave passes (Figure 5.8(b)). So, as with other types of wave motion, sound waves can be characterized by their amplitude (a measure of intensity or loudness of the sound) and frequency (f) or wavelength (λ, lambda), which are related to speed (c) by the expression:

$$c = f\lambda \qquad (5.7)$$

5.2.1 THE MAIN CHARACTERISTICS OF SOUND WAVES IN THE OCEANS

The wavelengths of acoustic energy that are of interest in the ocean range from about 50m to 1mm, which, taking the velocity of sound in seawater as approximately $1500\,\mathrm{m\,s^{-1}}$, corresponds to frequencies from 30Hz* to 1.5MHz. (For comparison, sound frequencies above about 20kHz cannot be heard by the normal human ear.)

When acoustic energy is emitted uniformly in all directions by a point source in the middle of a homogeneous mass of seawater, it spreads outwards, producing spherical surfaces of constant pressure (remember that these are compressional waves), centred on the point source. The acoustic intensity will decrease with increasing distance from the source as a result of:

1 **Spreading loss** due to being spread over an increasingly greater surface area. The surface area of a sphere is proportional to the square of the radius of the sphere, and thus the spreading loss is proportional to the

*The hertz (Hz) is the unit of frequency = 1 cycle per second. 1 kHz = 1 kilohertz = 10^3Hz. 1 MHz = 1 megahertz = 10^6Hz.

square of the distance travelled. Spreading loss is independent of frequency (see also Section 5.2.2). (Spherical spreading loss also occurs in the case of light, of course, but attenuation is so great over short distances that spreading loss is less important.)

2 **Attenuation** due to **absorption**, the conversion of acoustic energy into heat and chemical energy; and **scattering**, due to reflection by suspended particles and air bubbles. Scattering is largely independent of frequency; absorption is not. At high frequencies, viscous absorption predominates (i.e. absorption due to the viscosity of the water itself), and in freshwater this is the dominant cause of attenuation by absorption over much of the frequency range (Figure 5.9). However, in seawater at intermediate and low frequencies, the principal mechanism of absorption is dissociation of the $MgSO_4$ ion pair and of the $B(OH)_3$ complex (see Chapter 6). These split up into their constituent ions on the passage of a sound wave, and this process extracts energy from the sound wave—it is called 'relaxation' by acousticians. At frequencies of a few hundred Hz, it seems likely that the main cause of attenuation by absorption is inhomogeneities in the water column.

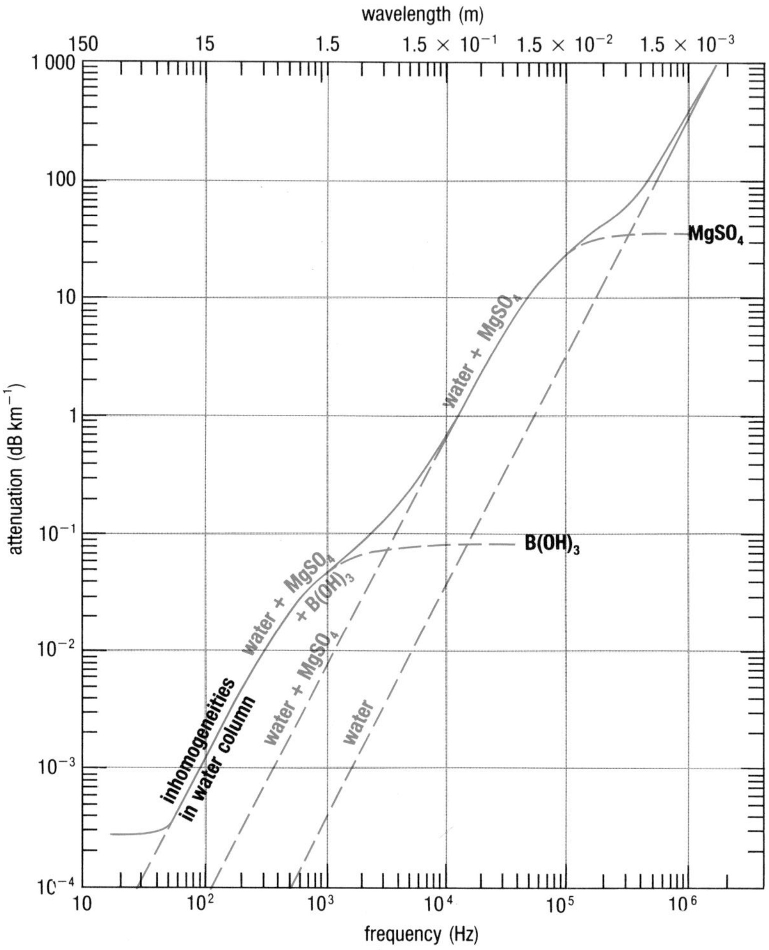

Figure 5.9 The attenuation of acoustic energy as a function of frequency in seawater, showing the dominant causes of attenuation and how they change according to frequency. The curves in this diagram are for a specific temperature and pressure; attenuation varies somewhat according to changing conditions. (dB = decibel, the unit of measurement for sound intensity.)

5.2.2 THE SPEED OF SOUND: REFRACTION AND SOUND CHANNELS

The speed, c, of compressional waves is given by

$$c = \sqrt{\frac{\text{axial modulus}}{\text{density}}} \tag{5.8}$$

The axial modulus of a material is a measure of its elasticity or compressibility, in other words, of the material's rigidity or 'stiffness', e.g. a billiard ball is more elastic than a tennis ball.

QUESTION 5.8 From equation 5.8, c varies inversely with density, implying that denser materials have lower acoustic velocities. This is seldom the case in naturally occurring materials. For instance, can you suggest how the speed of sound in water compares with that in air and in rock? Can you account for the apparent anomaly with the help of equation 5.8?

Both the axial modulus (elasticity or compressibility) and the density of seawater depend on its temperature, salinity and pressure, and thus c becomes rather a complex function of these three variables in the ocean.

Raising the temperature of seawater lowers its density, and thus, from equation 5.8 we should expect the speed of sound, c, to increase with increased water temperature. In the surface layers of the oceans, an increase in temperature of 1°C leads to an increase in c of about $3\,\mathrm{m\,s^{-1}}$.

We know that increased salinity leads to higher density, and so, at first sight, from equation 5.8 the speed of sound should decrease with increasing salinity. However, an increase in salinity also increases the axial modulus (the liquid becomes less compressible), and this counteracts the increase in density. For example, in surface layers of the oceans, an increase of 1 part per thousand in salinity actually results in an *increase* of about $1.1\,\mathrm{m\,s^{-1}}$ in c (the speed of sound in seawater is therefore greater than in freshwater—see Table 5.1).

Just as the speed of (seismic) sound waves increases with depth in the Earth, so the speed of acoustic waves increases with depth in the oceans (except in the sound channel, see below). The increase in axial modulus with depth is greater than the corresponding increase in density, and so c becomes greater (equation 5.8). An increase in depth of 100m will produce an increase in pressure of almost exactly 10 atmospheres ($10^6\,\mathrm{N\,m^{-2}}$, Figure 4.3), and the effect of this is to increase c by about $1.8\ \mathrm{m\ s^{-1}}$.

In the top few hundred metres of the ocean, where temperature changes are large (Figures 2.6–2.8), c will be controlled mainly by temperature, and to a much smaller degree by salinity and depth. Below the permanent thermocline, however, neither T nor S varies greatly, and so pressure becomes the dominant control on c.

A convenient empirical formula for the speed of sound in seawater over the temperature range 6°C to 17°C is:

$$c = 1\,410 + 4.21T - 0.037T^2 + 1.14S + 0.018d \tag{5.9}$$

where T and S are, of course, temperature and salinity, and d is depth (in metres), to which pressure is directly proportional.

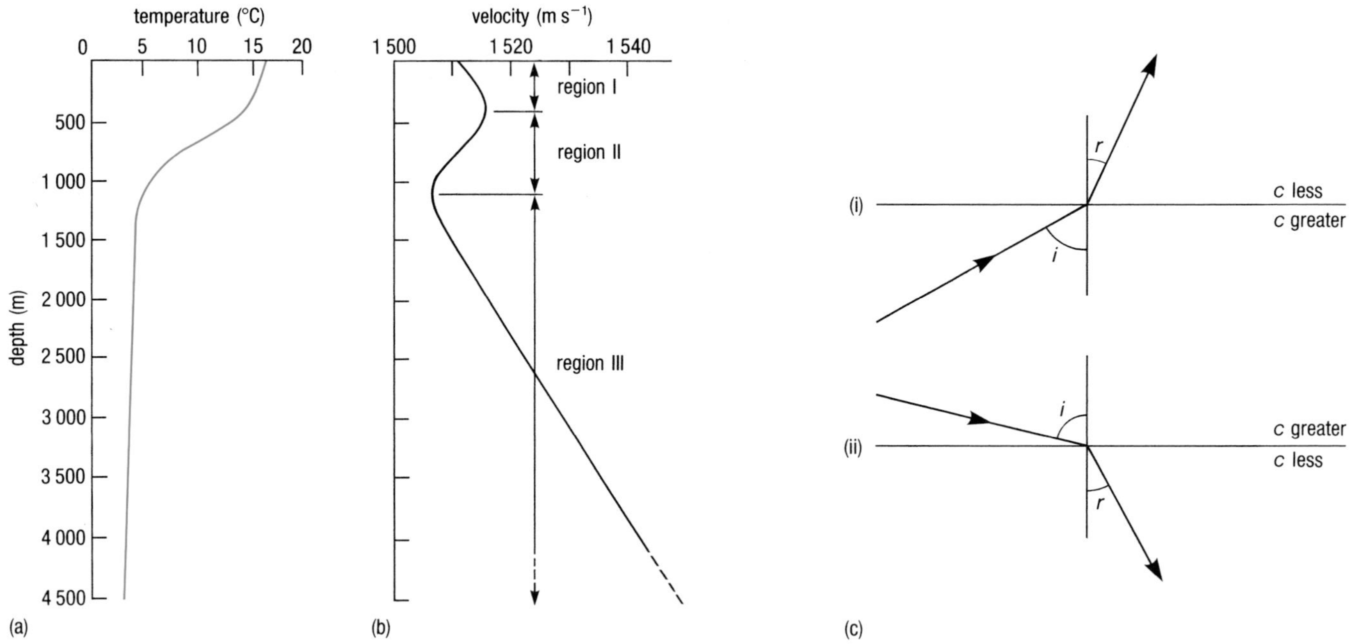

Figure 5.10 (a) A typical temperature–depth profile in the ocean.

(b) A typical profile of the speed of sound in the ocean.

(c) Idealized sketches illustrating refraction at interfaces where the speed of sound changes.
(i) Upward refraction (regions I and III); and
(ii) downward refraction (region II). From Snell's law:

$$\frac{c \text{ greater}}{c \text{ less}} = \frac{\sin i}{\sin r}$$

QUESTION 5.9 Estimate the speed of sound in seawater of temperature 10°C and salinity 35, at 100m depth.

Horizontal variations in *c* are very much smaller than vertical ones because horizontal gradients of temperature and salinity are much smaller than vertical gradients. Thus, an acoustic wave travelling vertically in the ocean will not be significantly affected by refraction because it is travelling essentially at right angles to the interfaces between layers of different density. However, a wave travelling horizontally may be subject to considerable refraction because it will meet such interfaces at low angles. In regions I and III of Figure 5.10(b), a sound wave will be refracted upwards, because the speed of sound decreases upwards (*cf.* Figure 5.10(c)) whereas in region II it will be refracted downwards, because the speed of sound decreases downwards (*cf.* Figure 5.10(c)).

The paths that would be followed by acoustic waves may be determined from a knowledge of the values of *c* throughout the ocean, and ray diagrams can be drawn, as in Figure 5.11. The rays are simply lines drawn perpendicular to the propagating wave front, and they therefore represent the direction of propagation. Note that most rays are focused on the boundary between regions II and III, whereas there is a **shadow zone** in the vicinity of the boundary between regions I and II that is penetrated only by waves that have been reflected at the surface of the ocean. The channel in which rays are trapped by refraction at the boundary between regions II and III is known as the **sound channel**.

The spreading loss for energy emitted in the sound channel is proportional only to the distance travelled. This is because the energy is constrained by the sound channel to spread outwards mainly in the two horizontal dimensions. Therefore, the surfaces of constant acoustic pressure are cylindrical, not spherical (*cf.* Section 5.2.1, item 1), and the surface of a cylinder is proportional to its radius (Figure 5.12). The information summarized in Figures 5.11 and 5.12 is of considerable significance in the use of acoustic energy in the oceans.

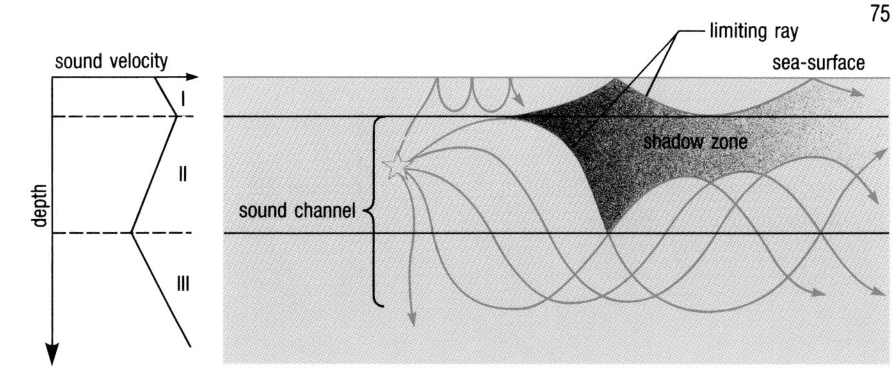

Figure 5.11 An example of a ray diagram for a sound emitted in region II of Figure 5.10 (b), showing a sound channel and a shadow zone. (See text for further discussion.) The shadow zone is defined by the limiting rays, reflected at the sea-surface and/or the boundary between regions I and II.

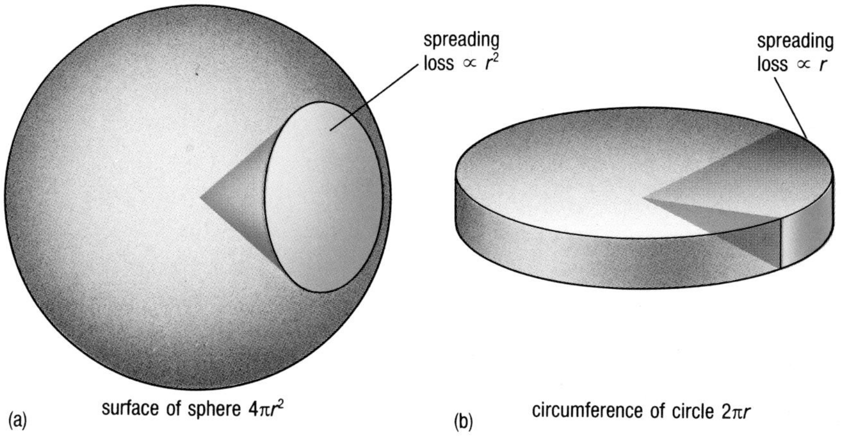

Figure 5.12 Illustrations showing:
(a) Spherical spreading loss from a point source. Surfaces of constant acoustic pressure are spherical and spreading loss is proportional to r^2.
(b) Cylindrical spreading loss from a point source, as in the sound channel. Surfaces of constant acoustic pressure are cylindrical and spreading loss is proportional only to r.

5.2.3 USES OF ACOUSTIC ENERGY IN THE OCEANS

The main disadvantage of using sound waves, in comparison with light waves, is their much greater wavelength (or lower frequency) which means that the resolution they can provide is much less; i.e., the smallest object that can be distinguished is comparatively large (about three wavelengths). Also important is the directional response of an emitting or receiving device. For example, to confine an acoustic beam to a width of 1° requires an emitting device some 60 wavelengths long. For a wavelength of 30mm (frequency about 50kHz), therefore, the required length would be 1.8m. The aims (of achieving the greatest possible resolution and reducing attenuation to achieve the maximum range) produce a conflict with regard to the wavelength of the acoustic energy to be used.

Why is there such a conflict?

Figure 5.9 shows that attenuation is very frequency-dependent, and this is of great importance in selecting operating frequencies for different uses and applications of underwater acoustic systems. For example, losses from attenuation are about 5% per nautical mile (3% per km) at 5kHz, increasing to about 90% per nautical mile (70% per km) at 30kHz. So, to keep attenuation to a minimum, the lowest possible frequency must be used. However, we have just seen that for maximum resolution the highest possible frequency is desirable. Designers of acoustic systems used in the oceans therefore have to arrive at a compromise, according to whether range or resolution is more important.

Applications of acoustic energy in the oceans

The ways in which acoustic waves are used in oceanic investigations may be divided into three major categories:

1 *Passive listening systems*: These involve the use of receiving devices—hydrophones—to listen to the sounds that are present, such as those emitted by whales, fish, or submarines. Analysis of the frequency spectra of the 'sounds' will usually assist in identifying their sources.

2 ***Sonar*** (*SOund Navigation And Ranging*): An acoustic signal is emitted and reflections are received from objects within the water (perhaps fish or submarines) or from the sea-bed. When the acoustic wave travels vertically down to the sea-bed and back, the time taken will provide a measure of the depth of the water, if *c* is also known (either from direct measurement or from temperature, salinity and pressure data). This is the principle of the echo-sounder, which is now universally used on sea-going vessels. A commercial echo-sounder may well have a beam width of 30–45° around the vertical, but for specialized applications (such as the detection of fish or submarines, or detailed studies of the sea-bed) beam widths of less than 5° are used and facilities provided for varying the direction of the beam. Note that although Figure 5.10 shows the effects of temperature, salinity and pressure on the speed of sound in seawater (*c*. 1 500 m s^{-1}) to be relatively small, even slight changes in *c* can lead to appreciable errors in measuring depth, and the degree of error may be increased by poor resolution.

Echo-sounding techniques for depth determination and sea-bed mapping have become very sophisticated, with the development of towed sonar devices, such as *SEABEAM*, a multi-beam echo-sounding system which determines the water depth along a swath of sea-floor beneath the towing ship, producing very detailed bathymetric maps. Sidescan imaging systems, such as *GLORIA* (*G*eological *LO*ng *R*ange *I*nclined *A*sdic) and *SeaMARC*, produce the equivalent of aerial photographs or radar images, using sound rather than light or microwaves to do so. Echo-sounding is also much used by fishermen, because even individual fish will produce an echo (see below), and shoals of fish or other animals can be identified as **scattering layers** within the water column (Figure 5.13).

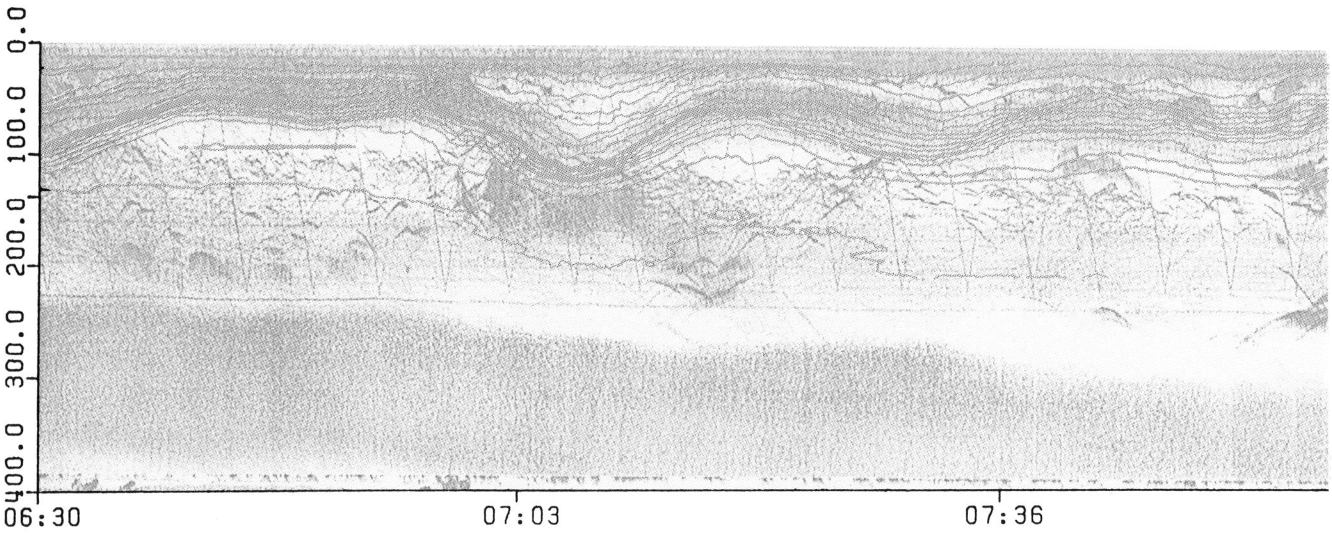

Figure 5.13 A typical sonogram (at 50 kHz) showing two scattering layers. The depth scale is in metres and horizontal scale gives the time of day. The undulating upper band is the thermocline, with the temperature structure (independently but simultaneously determined) superimposed as contours at 0.1 °C intervals. (The topmost contour is 10.9 °C.) Scattering in the thermocline could be due to backscatter caused by changes in acoustic impedance (see following text) associated with changes of temperature and density. Less coherent scattering below the thermocline is due to to fish and zooplankton. The lower and more regular scattering layer (250–300 m depth) is believed to be due to the zooplankton *Meganactyphanes norvegica*, a species which undertakes diurnal vertical migrations. The regular zig-zag line is the trace of a conductivity–temperature–density (CTD) probe being used in 'yo-yo' mode.

Sonar has well-known military applications, especially in submarine warfare; and many marine animals have sonar-type mechanisms for the echo-location of prey or of other individuals in the group, as well as for identification and communication. Whales and dolphins are perhaps the best known for this ability—whales have been known to communicate with one another across entire oceans, using the sound channel. It is said that dolphins are also capable of stunning or even killing their prey with bursts of very intense acoustic energy; and that squid and octopus have evolved a simple protection against this form of attack—they are deaf.

Acoustic impedance is a measure of the acoustic behaviour of a material and determines how good a 'target' it will be for sonar systems:

$$\text{impedance, } Z = \rho c \tag{5.10}$$

So the acoustic impedance of seawater is about:

$$1.03 \times 10^3 \text{ kg m}^{-3} \times 1\,500 \text{ m s}^{-1} = 1.55 \times 10^6 \text{ kg m}^{-2} \text{ s}^{-1}$$

The reflection of acoustic energy can only occur at the interface between two media of different acoustic impedances. For reflection normal to the interface, the reflectivity R, is given by:

$$R = \frac{Z_1 - Z_2}{Z_1 + Z_2} \times 100\% \tag{5.11}$$

where Z_1 and Z_2 are the acoustic impedances of the two materials on either side of the interface.

QUESTION 5.10 What is the reflectivity if $Z_1 = Z_2$?

Similarly, reflectivity will be at a maximum where $Z_1 - Z_2$ is greatest. Table 5.1 gives typical values of c, Z and R for some common materials.

Table 5.1 Acoustic properties of some common materials.

Material	Acoustic velocity c (m s^{-1})	Acoustic impedance $Z = \rho c$ ($\times 10^6$)	Reflectivity in seawater R(%)
Air (20°C)	343	415*	100
Freshwater (15°C)	1481	1.48	—
Seawater (35‰, 15°C)	1500	1.54	—
Wet fish flesh	~1450	1.6	1.9
Wet fish bone	~1700	2.5	2.4
Steel	6100	47	94
Brass	4700	40	92
Aluminium	6300	17	83
Perspex	2570	3.06	33
Rubber	1990	1.81	8
Concrete	3100	8	68
Granite	5925	16.0	82
Quartz	5750	15.3	82
Clay	~3000	7.7	67
Sandstone	~3300	~ 7.6	67
Basalt	~6000	~16.8	84

*The value given for air is *not* $\times 10^6$.

3 *Telemetry and tracking*: Locations may be identified and objects tracked in the oceans if they are equipped with acoustic transmitting devices. If they emit their signals in the sound channel, they can be monitored by hydrophones perhaps thousands of kilometres away. This is the basis of **Sofar** (*SO*und *F*ixing *A*nd *R*anging) technology, which has been developed considerably since World War II, and is widely used for military purposes, such as the location of submarines, wrecked aircraft and sunken ships. Scientific use includes locating the epicentres of submarine earthquakes and the charting of subsurface currents by means of floats equipped with acoustic sources. The density of these Sofar floats can be adjusted so that they are neutrally buoyant at a particular specified depth (i.e. they sink to that depth and stay there, because their density is the same as that of the surrounding water), and they then drift passively in the prevailing current at that depth.

In addition to tracking the movement of water in the currents which move them along, transmissions from Sofar floats can also be used to pass other information. For example, if a temperature-sensing device is arranged to control either the frequency of the transmitted signal or the interval between successive signals, the temperature of the surrounding water can be gauged.

The accuracy of Sofar fixes depends on reliable knowledge of the speed of sound in the oceans, especially within the sound channel. Figures 5.10 and 5.11 can be regarded as representing theoretical ideal situations. In practice, factors including seasonal and other fluctuations of temperature and salinity in time and space can lead to variations in the depth of the sound channel as well as losses from it. However, it is important to bear in mind that sound waves (rays) which 'leak' out of the sound channel tend to be reflected or refracted back into it (Figure 5.11): sound in the oceans travels with least loss in the sound channel, and Sofar devices work best within or near it. For this reason, the sound speed structure of the sound channel has been mapped in some detail over most of the oceans, both by direct measurement and by computation using equations such as equation 5.9 and the many thousands of T and S measurements that have been made over the years. Figure 5.14 is a section resulting from one such compilation.

QUESTION 5.11 (a) Is the speed of sound in the sound channel the same throughout Figure 5.14?

(b) What can you say about variations in the depth of the axis of the sound channel with latitude?

(c) Why does sound speed increase above the sound channel, where both temperature and salinity increase; and also below it, where temperature and salinity decrease?

Acoustic noise

When specific acoustic signals are being emitted and then listened to, as in applications 2 and 3 above, all other acoustic energy in the ocean is considered as **noise** above which the required signal is to be heard (this is analogous to the way in which 'atmospherics' can swamp the weak signal from a distant radio source). Amplifying a weak signal in an attempt to enable it to be heard above the noise simply increases both signal and backgroud noise; in addition, **reverberations** (multiple reflections by particles in the water as well as at the ocean boundaries) may become a

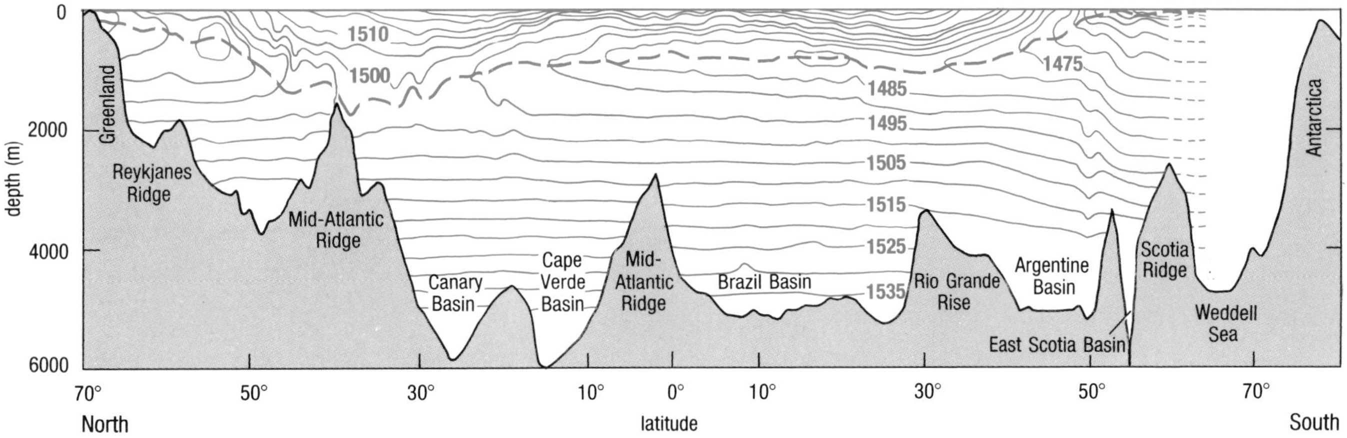

Figure 5.14 North–south section of sound channel structure in the north and south Atlantic along the 30.5° W meridian. Sound speeds are in m s^{-1}, and the approximate sound channel axis is indicated by a heavy blue broken line. Contours of equal speed are based on annual average data. Near-surface structure above the axis in mid-latitudes is subject to seasonal variations. These average contours are included only as a general indication of the complete vertical structure. Sound speeds generally increase from the sound channel axis to the base of the mixed surface layer.

severe problem. Some noise in an acoustic system may be caused by its own electrical circuits or the electrical system of the ship; the ship may also be the source of mechanical noise, from engines and other equipment.

Ambient noise produced in the sea itself falls into two categories: physical and biological. Physical noise is mostly wind-induced and is in the audible frequency range (between about 10 and 10^4 Hz): it includes the sound of waves and bursting bubbles, rainfall, moving ice and sediments shifting on the sea-bed. Biological noise is produced by communicating whales and dolphins, by the activity of some Crustacea (e.g. snapping shrimps), and by certain fish.

Most of the biological noise produced by marine animals is at very low frequencies, less than 50Hz, and most animal detectors work in that range—the lateral line system of many fishes, for example, is highly sensitive to these low frequencies. Only those animals with specialized auditory receptors can use sound for communication—in some deep-living fish, those receptors are the swim-bladders that are normally used for buoyancy.

5.2.4 ACOUSTIC OCEANOGRAPHY

Since the early 1970s there have been considerable advances in the application of acoustic techniques to the investigation of relatively small scale and short term vertical and lateral inhomogeneities within and between water masses, including microstructure and fronts as well as other features. There are two categories of approach in this kind of research.

Attenuation experiments

Figure 5.9 shows that at low frequencies, the main cause of attenuation of sound in the sea is inhomogeneities in the water column. Different water masses have different patterns of inhomogeneity. As T and S and other properties change from one water mass to another, so does the degree of attenuation. Long-range, low-frequency sound transmissions using the sound channel thus provide an additional means of identifying boundaries between different oceanic water masses.

What is another reason for using low frequencies?

These are long-range experiments, and attenuation is strongly frequency-dependent; so the lower the frequency, the greater the range.

Acoustic tomography

A major purpose of this more ambitious technique is to identify and monitor the progress of what are known as *meso-scale eddy systems*. These bear the same relationship to oceanic water masses (Section 4.1) that atmospheric depressions and anticyclones have to the major air masses, but are about ten times smaller.

The existence of meso-scale eddies was not even suspected until the 1960s and was not demonstrated until the 1970s, so this whole field of research is relatively new. The eddies have length scales of the order of 100km and time-scales ('lifetimes') of the order of several months, and they are difficult both to detect and to track using conventional T, S and current-measuring equipment. The principle of acoustic tomography is very simple, but the actual method and attendant computation is extremely complex. The method relies on the fact that individual eddies have temperatures different from those in the surrounding water—there are both warm and cold eddies. It follows that the speed of sound between an acoustic source and receiver will change if an eddy passes between them.

In practice, the 'travel time' of sound between source and receiver is measured. Would the travel time increase or decrease if a cold eddy passed through?

A fall in temperature causes a decrease in the speed of sound—so the travel time would increase. In a typical experiment, a whole array of moored acoustic sources and receivers is used to monitor a 'volume' of ocean that may be from 300–1000km across (Figure 5.15(a)). Acoustic pulses are emitted every few seconds, at a frequency of about 250Hz, and sound travel times (about 11 minutes for 1000km) are measured between each transmitter/receiver pair. The data set is enormous and analysis of the results requires powerful computers. Figure 5.15(b) gives one set of sound ray paths for just one source/receiver pair, and provides an idea of the degree of complexity that must be unravelled in order to chart the passing of an eddy system.

QUESTION 5.12 (a) Temperature changes of a degree or two and salinity changes of 0.1 are typically encountered across eddy boundaries. How important is salinity in the context of acoustic tomography?

(b) Figure 5.15(b) shows that sound rays travelling within the sound channel can actually arrive *later* than rays which have been down to the bottom and then are refracted back up again. Why?

Conventional ship-based measurements of temperature and current velocity are also made in the area of interest. Acoustic travel times are

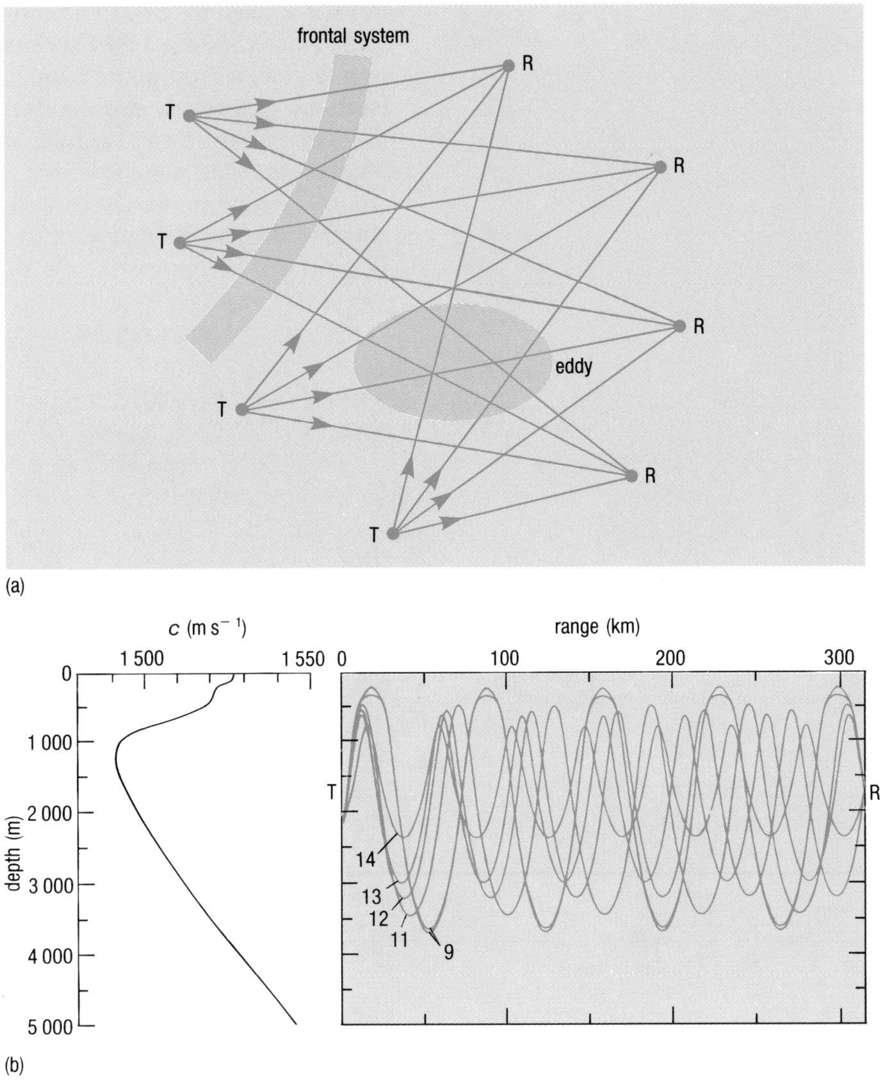

Figure 5.15 (a) Sketch plan of acoustic tomography array, showing the multiple transmission paths between acoustic transmitters (T) and receivers (R). Note that while we concentrate on the application of the technique to locating meso-scale eddies, other features such as fronts can also be detected.

(b) A multipath ray diagram between one transmitter (T) and one receiver (R) in an acoustic tomography experiment in the north-western Atlantic. Only the ray paths for acoustic pulses that initially travel *upwards* are shown. An equal number of initially downward-directed ray paths would give a more or less mirror-image pattern of lines on the diagram, but these have been omitted for clarity. The numbers give the total number of 'turning points' for each ray. Note that the deeper the ray, the fewer these are. The depth profile for the average speed of sound for this transmitter/receiver pair is shown in blue on the left.

affected not only by the properties of the water through which the sound travels, but also by the currents transporting that water. Clearly, currents travelling with the sound will reduce travel times, and those travelling against it will increase travel times.

Physical oceanographers use all these measurements along with the acoustic data to make mathematical models of the temperature structure of the region being investigated. It is an iterative process, using what is known as inverse theory: the theoretical models are compared with the actual measurements, moving by successively closer approximations towards a 'best fit' between theory and observation. Figure 5.16 shows some results from a 1981 experiment and demonstrates how acoustic tomography was combined with conventional ship-based measurements. Both the technology and the computational techniques are being improved all the time, enabling more thorough and rigorous analysis of the data to make more detailed maps of oceanic meso-scale circulation. An experiment in the Greenland Sea was begun in the late 1980s to detect and monitor the formation of deep water masses.

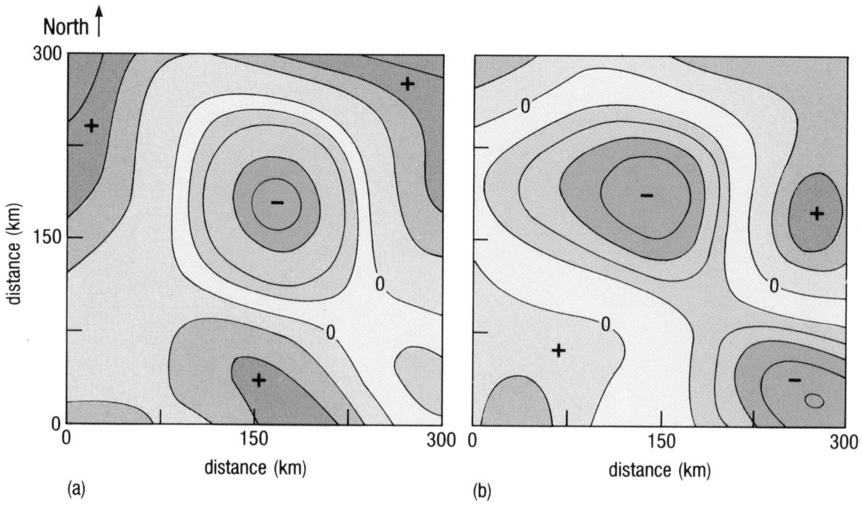

Figure 5.16 Progress of eddies mapped in the 1981 experiment in the north-western Atlantic. Contours represent equal sound speed at 700 m depth, relative to a reference speed. Negative regions (blue) represent lower sound speeds in cold water; positive regions (red) represent higher sound speeds in warm water. The cold eddies in the centre and south-east of the area (a) begin to spread and move westwards (b) as a warm eddy enters from the east.
(a) is from ship surveys, (b) is from acoustic tomography.

Although they are not intrinsic properties of seawater, the behaviour of light and sound in the oceans has enormous practical implications; and to understand this properly does require a thorough knowledge of the intrinsic properties themselves. In the next Chapter, we look at the behaviour of the various constituents that help to give seawater its many interesting characteristics.

5.3 SUMMARY OF CHAPTER 5

1 Light and all other forms of electromagnetic radiation travel at a speed of 3×10^8 m s^{-1} in a vacuum (about 2.2×10^8 m s^{-1} in seawater). Light propagated through water is subject to absorption and scattering, and its intensity decreases exponentially with distance from the source. Sunlight sufficient for photosynthesis cannot penetrate to more than about 200m depth, and this defines the limit of the photic (or euphotic) zone, within which photosynthetic primary production can occur.

The aphotic zone extends from the bottom of the photic zone to the sea-bed. Sunlight penetrates through only about the upper 1000m of the aphotic zone; below that, the oceans are permanently dark. Light intensities under water decrease exponentially with distance from the surface or other light source, as a result of attenuation by absorption and scattering. The downwelling irradiance from sunlight or moonlight provides the non-directional (diffuse) light required for underwater illumination. Underwater vision requires directional light: light must travel direct from object to eye for a coherent image to be formed. Directional light is subject to greater attenuation than non-directional light.

2 Underwater visibility depends on contrast, which is a function partly of object brightness or reflectivity and partly of attenuation with distance. Below depths of a few tens of metres, underwater light becomes virtually monochromatic, so contrast is mostly a matter of differences of light intensity rather than of colour. In lower parts of the aphotic zone, where many fish have bioluminescent organs (photophores), light is used in the same way as colour is used on land—for inter- and intraspecific recognition, camouflage, deterring predators, and so on.

3 Beam transmissometers and turbidity meters are used to determine the attenuation coefficient (C) of directional light, and irradiance meters are used to determine the diffuse attenuation coefficient (K) of the non-directional downwelling irradiance. Nephelometers measure scattering and can be used to determine concentrations of particulate matter in the water. The Secchi disc is a simple piece of equipment for measurement of underwater light. By applying simple empirical equations, the measurements can be used to estimate visibility, attenuation coefficients, and depth of the photic zone.

4 Water preferentially absorbs longer wavelengths of the electromagnetic spectrum, which is why water appears as blue. Yellow substances and suspended particles absorb shorter wavelengths, so turbid water tends to look yellow in colour, while productive ocean waters have the green colour of chlorophyll. In clear water, about 35% of incident blue–green light penetrates to 10m depth. In turbid water, about 2% of yellow–green light penetrates to 10m depth. Photosynthesis is considerably inhibited in turbid waters.

5 Passive remote sensing of the oceans makes use of reflected and radiated visible, infrared and microwave radiation, to determine properties such as sea-surface temperature and water colour. Active remote sensing uses microwave imaging radar techniques to obtain information about the state of the sea-surface. Electromagnetic radiation cannot penetrate far through water, so remote sensing with the electromagnetic spectrum can provide direct information only about the sea-surface; and radio communication is all but impossible below the surface of the ocean.

6 Sound travels much more slowly than light through water but can travel much further, and so is used for remote sensing and communication in the oceans. Frequencies of interest in the oceans lie approximately in the 30Hz to 1.5MHz range. Sound intensities decrease with distance from the source because of two processes. (a) Spreading loss, due to being spread out over (i) the surface of a sphere (loss proportional to distance^2), or (ii) the surface of a cylinder (loss proportional to distance), as in the sound channel, see below. (b) Attenuation, due to (i) absorption by the water and reactions involving its dissolved constituents, notably the dissociation of $B(OH)_3$ and $MgSO_4$; (attenuation increases as frequency increases, and high frequencies are very rapidly attenuated); and (ii) scattering, i.e. reflection by suspended particles.

7 The speed of sound in seawater, c, increases as the axial modulus of seawater increases, and decreases as the density increases; it is about 1 500 m s^{-1}. A temperature rise of 1 °C causes an increase of about 3ms^{-1}. A salinity increase of 1 causes an increase of about 1.1ms^{-1}. A pressure increase equivalent to an increase in depth of 100m causes an increase of about 1.8ms^{-1}. The speed of sound is at a minimum both at the surface and in the sound channel.

8 Sonar is used for depth determination, sea-bed mapping, and the location of objects, especially fish and submarines; and many marine animals also make use of the technique. The reliability of echo-sounding depends partly upon the acoustic impedance: the higher the impedance contrast between water and the material of the object sought, the better the 'target' provided.

9 Sofar is used for longer-range location, and also for tracking, especially of neutrally buoyant acoustic floats within and near the sound channel. To fix the position of Sofar devices reliably, variations of the speed of sound throughout the oceans must be known as accurately as possible. The axis of the sound channel lies between about 0.5 and 1.5km depth throughout most of the oceans, between the latitudes of 60°N and S. Above these latitudes there is no sound channel.

10 In any acoustic receiving system there is background noise due to ambient sounds emanating from instrumental, physical and biological sources; and reverberation due to multiple reflections—scattering—by particles.

11 Acoustic oceanography is a relatively new field of research, involving two main approaches. Attenuation experiments depend on the fact that, over long distances, variations of T and S and inhomogeneities in the water column attenuate low frequencies in the sound channel to different extents, and this allows the boundaries of water masses to be located. Acoustic tomography depends on the fact that the speed of sound varies according to temperature, and—with the help of mathematical modelling techniques—is used to detect and monitor warm and cold eddy systems that form a part of the oceanic circulation.

Now try the following questions to consolidate your understanding of this Chapter.

QUESTION 5.13 Why are the longer wavelengths of the spectrum missing on Figure 5.7?

QUESTION 5.14 Some fishes are better acoustic targets than suggested by the reflectivity of wet fish flesh in Table 5.1. Can you explain why this might be?

QUESTION 5.15 Would you select a high or low frequency sonar system for (a) fishing, (b) submarine detection?

QUESTION 5.16 Which of the following statements are true?

(a) The depth of the photic zone is typically much greater near coasts than in the open ocean.

(b) The waters of the Gulf Stream issuing from the Straits of Florida are highly productive (see quotation in Section 4.4.3).

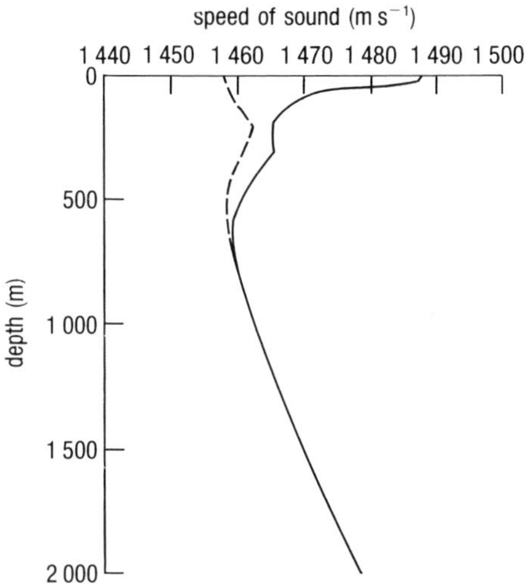

Figure 5.17 Variation of the speed of sound with depth at two seasons of the year. For use with Question 5.17.

(c) Equation 5.9 could not in general be used for water below the thermocline.

(d) The reflectivity of the air–sea interface in Table 5.1 should more accurately be given as 99.946%.

QUESTION 5.17 Examine Figure 5.17. Explain which curve is for the winter, and which is for the summer. What is the depth of the sound channel axis, and why are seasonal changes minimal at and below this depth?

CHAPTER 6 THE SEAWATER SOLUTION

'In my first science lesson we all watched intently round a pan of clear boiling salt water. When all the water had evaporated, we were awed at the sight of salt left at the bottom of the pan. Where had it come from?'
(Viki Capel, *New Scientist*, 1987.)

Up to now we have mainly treated the salinity of seawater as a quantity approximating to 35 parts per thousand by weight. However, you have been introduced to the major ions that contribute 99.9 per cent of the salinity (Table 3.1) and in Section 3.1 we established the important principle of the constancy of composition of seawater. In this Chapter, we consider seawater as an aqueous solution and examine the sources and the behaviour of some individual constituents.

6.1 THE GROSS CHEMICAL COMPOSITION OF SEAWATER

Most of the 92 naturally occurring elements have been measured or detected in seawater, and the remainder are likely to be found as more sensitive analytical techniques become available. The elements so far determined show a vast range of concentrations, as you can see from Table 6.1.

QUESTION 6.1 There are major differences in the figures for sulphur, carbon and boron in Table 6.1, compared with Table 3.1. Why is that, and how might such disparities be avoided?

Particulate matter
There is a wide variety of suspended particles in seawater (the seston, Section 5.1), and the distinction between what constitutes material truly in solution and what is particulate matter can present problems in the determination of the concentrations of some elements in seawater. A widely used procedure for separating dissolved from particulate fractions is filtration through a membrane having pores of diameter 0.45μm. For most dissolved constituents this provides a satisfactory separation between dissolved and particulate matter, but for some it is less clear-cut. For example, iron in seawater occurs in hydrated forms such as $Fe(OH)_2$ or $Fe(OH)_3$. These tend to coalesce to form **colloidal** particles, which are so small that they remain in suspension indefinitely, unless some process occurs to aggregate them into particles large enough to settle under gravity. Thus, for iron there is a spectrum of sizes, ranging from true solution, through colloidal particles, to aggregated particles. Use of a membrane having pores of diameter 0.45μm therefore effects a purely arbitrary separation between dissolved and particulate fractions. The measured ratio of dissolved to particulate iron in a given sample can be increased or decreased simply by changing the pore size of the membrane, or by increasing the filtration pressure, which may break up the aggregates mechanically. This problem does not arise with all elements that occur in hydrated forms, however. In the case of $Al(OH)_3$

Table 6.1 Abundances of chemical elements in seawater.

Element		Concentration ($mg l^{-1}$) (i.e. parts per million, p.p.m.)	Some probable dissolved species	Total amount in the oceans (tonnes)
chlorine	Cl	1.95×10^4	Cl^-	2.57×10^{16}
sodium	Na	1.077×10^4	Na^+	1.42×10^{16}
magnesium	Mg	1.290×10^3	Mg^{2+}	1.71×10^{15}
sulphur	S	9.05×10^2	SO_4^{2-}, $NaSO_4^-$	1.2×10^{15}
calcium	Ca	4.12×10^2	Ca^{2+}	5.45×10^{14}
potassium	K	3.80×10^2	K^+	5.02×10^{14}
bromine	Br	67	Br^-	8.86×10^{13}
carbon	C	28	HCO_3^-, CO_3^{2-}, CO_2	3.7×10^{13}
nitrogen	N	11.5	N_2 gas, NO_3^-, NH_4^+	1.5×10^{13}
strontium	Sr	8	Sr^{2+}	1.06×10^{13}
oxygen	O	6	O_2 gas	7.93×10^{12}
boron	B	4.4	$B(OH)_3$, $B(OH)_4^-$, $H_2BO_3^-$	5.82×10^{12}
silicon	Si	2	$Si(OH)_4$	2.64×10^{12}
fluorine	F	1.3	F^-, MgF^+	1.72×10^{12}
argon	Ar	0.43	Ar gas	5.68×10^{11}
lithium	Li	0.18	Li^+	2.38×10^{11}
rubidium	Rb	0.12	Rb^+	1.59×10^{11}
phosphorus	P	6×10^{-2}	HPO_4^{2-}, PO_4^{3-}, $H_2PO_4^-$	7.93×10^{10}
iodine	I	6×10^{-2}	IO_3^-, I^-	7.93×10^{10}
barium	Ba	2×10^{-2}	Ba^{2+}	2.64×10^{10}
molybdenum	Mo	1×10^{-2}	MoO_4^{2-}	1.32×10^{10}
arsenic	As	3.7×10^{-3}	$HAsO_4^{2-}$, $H_2AsO_4^-$	4.89×10^9
uranium	U	3.2×10^{-3}	$UO_2(CO_3)_2^{4-}$	4.23×10^9
vanadium	V	2.5×10^{-3}	$H_2VO_4^-$, HVO_4^{2-}	3.31×10^9
aluminium	Al	2×10^{-3}	$Al(OH)_4^-$	2.64×10^9
iron	Fe	2×10^{-3}	$Fe(OH)_2^+$, $Fe(OH)_4^-$	2.64×10^9
nickel	Ni	1.7×10^{-3}	Ni^{2+}	2.25×10^9
titanium	Ti	1×10^{-3}	$Ti(OH)_4$	1.32×10^9
zinc	Zn	5×10^{-4}	$ZnOH^+$, Zn^{2+}, $ZnCO_3$	6.61×10^8
caesium	Cs	4×10^{-4}	Cs^+	5.29×10^8
chromium	Cr	3×10^{-4}	$Cr(OH)_3$, CrO_4^{2-}	3.97×10^8
antimony	Sb	2.4×10^{-4}	$Sb(OH)_6^-$	3.17×10^8
manganese	Mn	2×10^{-4}	Mn^{2+}, $MnCl^+$	2.64×10^8
krypton	Kr	2×10^{-4}	Kr gas	2.64×10^8
selenium	Se	2×10^{-4}	SeO_3^{2-}	2.64×10^8
neon	Ne	1.2×10^{-4}	Ne gas	1.59×10^8
cadmium	Cd	1×10^{-4}	$CdCl_2$	1.32×10^8
copper	Cu	1×10^{-4}	$CuCO_3$, $CuOH^+$	1.32×10^8
tungsten	W	1×10^{-4}	WO_4^{2-}	1.32×10^8
germanium	Ge	5×10^{-5}	$Ge(OH)_4$	6.61×10^7
xenon	Xe	5×10^{-5}	Xe gas	6.61×10^7
mercury	Hg	3×10^{-5}	$HgCl_4^{2-}$, $HgCl_2$	3.97×10^7
zirconium	Zr	3×10^{-5}	$Zr(OH)_4$	3.97×10^7
bismuth	Bi	2×10^{-5}	BiO^+, $Bi(OH)_2^+$	2.64×10^7
niobium	Nb	1×10^{-5}	not known	1.32×10^7
tin	Sn	1×10^{-5}	$SnO(OH)_3^-$	1.32×10^7
thallium	Tl	1×10^{-5}	Tl^+	1.32×10^7
thorium	Th	1×10^{-5}	$Th(OH)_4$	1.32×10^7

Table 6.1 (continued)

Element		Concentration (mgl^{-1}) (i.e. parts per million, p.p.m.)	Some probable dissolved species	Total amount in the oceans (tonnes)
hafnium	Hf	7×10^{-6}	not known	9.25×10^{6}
helium	He	6.8×10^{-6}	He gas	8.99×10^{6}
beryllium	Be	5.6×10^{-6}	$BeOH^{+}$	7.40×10^{6}
gold	Au	4×10^{-6}	$AuCl_2^{-}$	5.29×10^{6}
rhenium	Re	4×10^{-6}	ReO_4^{-}	5.29×10^{6}
cobalt	Co	3×10^{-6}	Co^{2+}	3.97×10^{6}
lanthanum	La	3×10^{-6}	$La(OH)_3$	3.97×10^{6}
neodymium	Nd	3×10^{-6}	$Nd(OH)_3$	3.97×10^{6}
silver	Ag	2×10^{-6}	$AgCl_2^{-}$	2.64×10^{6}
tantalum	Ta	2×10^{-6}	not known	2.64×10^{6}
gallium	Ga	2×10^{-6}	$Ga(OH)_4^{-}$	2.64×10^{6}
yttrium	Y	1.3×10^{-6}	$Y(OH)_3$	1.73×10^{6}
cerium	Ce	1×10^{-6}	$Ce(OH)_3$	1.32×10^{6}
dysprosium	Dy	9×10^{-7}	$Dy(OH)_3$	1.19×10^{6}
erbium	Er	8×10^{-7}	$Er(OH)_3$	1.06×10^{6}
ytterbium	Yb	8×10^{-7}	$Yb(OH)_3$	1.06×10^{6}
gadolinium	Gd	7×10^{-7}	$Gd(OH)_3$	9.25×10^{5}
praseodymium	Pr	6×10^{-7}	$Pr(OH)_3$	7.93×10^{5}
scandium	Sc	6×10^{-7}	$Sc(OH)_3$	7.93×10^{5}
lead	Pb	5×10^{-7}	$PbCO_3, Pb(CO_3)_2^{2-}$	6.61×10^{5}
holmium	Ho	2×10^{-7}	$Ho(OH)_3$	2.64×10^{5}
lutetium	Lu	2×10^{-7}	$Lu(OH)$	2.64×10^{5}
thulium	Tm	2×10^{-7}	$Tm(OH)_3$	2.64×10^{5}
indium	In	1×10^{-7}	$In(OH)_2^{+}$	1.32×10^{5}
terbium	Tb	1×10^{-7}	$Tb(OH)_3$	1.32×10^{5}
tellurium	Te	1×10^{-7}	$Te(OH)_6$	1.32×10^{5}
samarium	Sm	5×10^{-8}	$Sm(OH)_3$	6.61×10^{4}
europium	Eu	1×10^{-8}	$Eu(OH)_3$	1.32×10^{4}
radium	Ra	7×10^{-11}	Ra^{2+}	92.5
protactinium	Pa	5×10^{-11}	not known	66.1
radon	Rn	6×10^{-16}	Rn gas	7.93×10^{-4}
polonium	Po		$PoO_3^{2-}, Po(OH)_2$?	

IMPORTANT NOTES

1 Table 6.1 does not represent the last word on seawater composition. Even for the more abundant constituents, compilations from different sources differ in detail (*cf.* Note to Table 3.1). For the rarer elements, many of the entries in Table 6.1 will be subject to revision, as analytical methods improve. Moreover, most constituents behave non-conservatively (Section 4.3.4), making averages less meaningful.

2 Concentrations in Table 6.1 are by weight (mg l^{-1} or p.p.m.). While this is convenient for some purposes, for many others it is more useful to express concentrations in molar terms. There are 6×10^{23} atoms in a mole (Avogadro's number). One mole of any element (or compound) has a mass in grams equal to the atomic (or molecular) mass of the element (or compound). Thus, a mole of calcium weighs 40g; a mole of magnesium weighs 24g; a mole of carbonate ion (CO_3^{2-}) weighs $12 + (16 \times 3) = 60$ g; and so on.

and $Si(OH)_4$, for example, filtration can satisfactorily distinguish dissolved from particulate fractions.

The density of particulate matter is typically greater than that of seawater, so it tends to sink. However, the small size of most particles means that they can remain in suspension almost indefinitely.

The classical equation for the settling velocity, v, of a spherical object in a fluid medium is (Stokes' law):

$$v = \frac{1}{18} \times g \times \frac{\rho_1 - \rho_2}{\mu} \times d^2 \qquad (6.1)$$

where: g is the gravitational acceleration ($m\,s^{-2}$)
d is the diameter of the particle (m)
ρ_1 is the density of the particle ($kg\,m^{-3}$)
ρ_2 is the density of the fluid ($kg\,m^{-3}$)
μ is the molecular viscosity of the fluid ($N\,s\,m^{-2}$) (*cf.* Table 1.1)
and the velocity, v will be in $m\,s^{-1}$.

Equation 6.1 gives a first-approximation value for the speed at which seston particles sink in seawater.

QUESTION 6.2 The majority of particles making up the seston have diameters less than 2μm. Assume an average density 1.5 times that of seawater, take values of the density and viscosity of water from Tables 1.1 and 1.2, and use $9.8ms^{-2}$ for g. Use equation 6.1 to calculate v for a particle that is 2μm in diameter. About how long would such particle take to sink 1m?

The long period of time you calculated in Question 6.2 would be even greater if the particle were not spherical. In addition, it would be considerably extended by turbulence in the water column, which counteracts the tendency of particles to settle out of suspension. In order for sedimentary particles to reach the sea-bed from the surface in a reasonable period of time (say a month), they must be much larger than the size range typical of the seston. (In fact, equation 6.1 is valid only for spherical particles with diameters less than about 100μm. For particles with diameters greater than about 2mm, the settling velocity is proportional to $d^{1/2}$ and equation 6.1 has a different form. For particles in the size range 100μm to 2mm, settling velocity is proportional to d^n, where $2>n>1/2$.)

The major sources of particulate material in the oceans are:

1 Rivers, carrying particles in suspension to the sea, where the coarser fractions are deposited as sands, silts and clays.

2 Wind-borne dust (e.g. quartz grains, silicified grass cells, freshwater diatom skeletons, as well as organic detritus); in addition, volcanic ash particles and material derived from the break up of meteorites are continually supplied to the oceans by atmospheric fall-out. Much of this input rapidly sinks to the sea-bed, but some of the particles are small enough to contribute to the seston.

3 Biogenic particulate matter, i.e. particles resulting from primary and secondary biological production, comprising skeletal remains, faecal pellets and dead plant and animal matter (detritus); much of this material has particle sizes of 100μm or more, and sinks relatively quickly.

6.1.1 CLASSIFICATION OF DISSOLVED CONSTITUENTS

Major constituents of seawater are those that occur in concentrations greater than about 1 part per million (1×10^{-6}) by weight, and which account for over 99.9% of the dissolved salts in the oceans. The major constituents of seawater are conventionally taken to be those listed in Table 3.1. Despite their relatively high concentrations, nitrogen and oxygen and silicon (Table 6.1) are not generally included, because the first two are dissolved gases and the third is a nutrient; oxygen and silicon in particular are strongly non-conservative.

Minor and trace constituents make up the remainder of the elements in seawater. Although the distinction between the two is somewhat ill-defined, trace constituents are those with concentrations of about 1 part per billion (1×10^{-9}) by weight, or less. On that basis, elements below about titanium in Table 6.1 are trace constituents.

Apart from the obvious distinction between major constituents and the rest in terms of mass, there is another reason for distinguishing between them. As you should recall from Section 3.1, the oceanic distribution of individual major constituents is in general closely related to that of total salinity, because of the constancy of composition of seawater—these constituents mostly behave conservatively (Section 4.3.4). Minor and trace constituents on the other hand, generally behave non-conservatively, being affected by biological and chemical processes in which they are added to or removed from solution.

6.1.2 THE NUTRIENTS

Carbon is a fundamental requirement for the support of life anywhere on Earth. Because of the predominance of carbon dioxide among the dissolved gases (see Section 6.1.3), carbon forms the eighth most abundant dissolved element in ocean water (Table 6.1). The availability of dissolved carbon is therefore not a limiting factor in biological production. More important constraints are the intensity of illumination, supply of oxygen and the availability of nutrients, especially fixed nitrogen, chiefly as nitrate (NO_3^-), phosphorus as phosphate (PO_4^{3-}) and silicon as silica (SiO_2). They are utilized by phytoplankton—plant cells ranging in size from one to a few hundred microns—which drift in the surface waters of the oceans and photosynthesize carbohydrates from carbon dioxide and water.

Why can phytoplankton not grow below depths of about 100–200m?

Light is essential for photosynthesis, and you have seen in Section 5.1 that phytoplankton can grow only in the photic zone, which is rarely deeper than 200m and generally much less. It is in the photic zone, therefore, that nutrients are most heavily utilized. Phytoplankton form the base of oceanic food chains and nutrients move along the chains as grazing and predation take place. They are recycled within the water column by excretion and microbial breakdown of organic particulate matter (detritus). Sinking of larger biogenic particles (faeces and corpses) and the vertical movements of zooplankton and other animals feeding on phytoplankton and detritus combine to cause a progressive downward movement of nutrients out of the photic zone. As a result, typical concentration–depth profiles for nitrate and phosphate look like Figure 6.1(a) and (b).

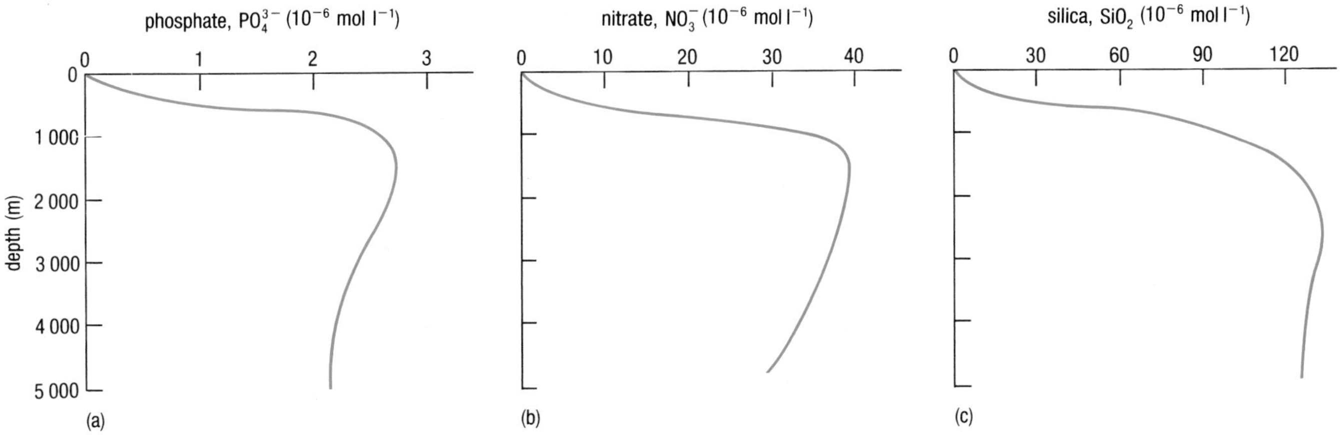

Figure 6.1 Typical concentration-depth profiles for (a) phosphate, (b) nitrate, and (c) silica. Note that concentrations are in mol l^{-1}.

The photic zone is continually being depleted of nutrients, and photosynthetic primary production will be inhibited unless there is vertical mixing, or vertical advection of nutrient-rich water from greater depths (upwelling). The term **biolimiting constituents** is sometimes applied to those nutrients whose availability in surface waters limits biological production. They include nitrate, phosphate and silica, and their characteristic vertical profiles, showing almost total depletion in surface waters (*cf.* Figure 6.1), are controlled principally by biological processes.

Nitrate and phosphate are used to form the soft tissues of organisms and the molar ratio of nitrate to phosphate in ocean water is close to the ratio of 15:1 for organic tissues (*cf.* Figure 6.1); thus, when all the dissolved phosphate in surface waters has been used up, so has all the dissolved nitrate. Why nitrate and phosphate should occur in seawater in the same ratio that organisms require them remains one of the intriguing mysteries of seawater chemistry. There is still no answer to the question of whether organisms evolved to use the 15:1 molar ratio of N:P because it was there, or whether marine organisms themselves established the ratio through time.

Nitrogen and nitrate

At this point we must make a short digression to clarify something that could cause confusion. Dissolved nitrogen gas as N_2 is used hardly at all in marine biological processes, because only minute amounts of it are fixed (i.e. incorporated into living tissue) by bacteria. Average oceanic water contains about 9 ml l^{-1} (parts per thousand by volume) of nitrogen gas (see Section 6.1.3 and Figure 6.4), which is equivalent to about 11 mg l^{-1} (p.p.m. by weight). The total concentration of nitrogen in average ocean water is given as 11.5 p.p.m. in Table 6.1; so, only a very small fraction can be in a form other than N_2 gas dissolved from the atmosphere. This is the fixed (i.e. chemically combined) nitrogen supplied to the oceans by rivers, mainly as nitrate produced by terrestrial nitrogen-fixing bacteria.

In fact, the average concentration of dissolved nitrogen as nitrate in seawater is generally of the order of 0.5 p.p.m. (0.5 mg l^{-1}). In molar terms, this agrees with the values in Figure 6.1 (a).

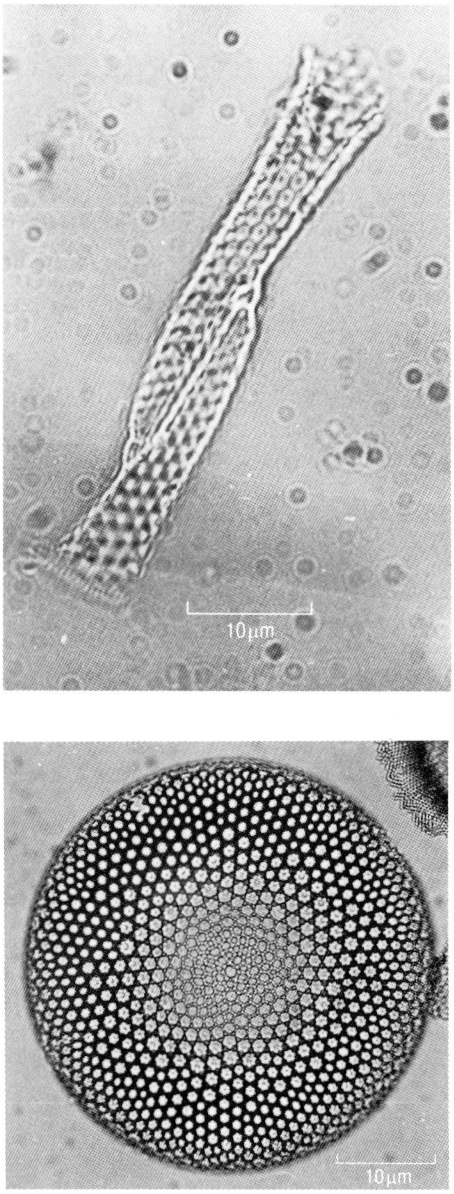

(a)

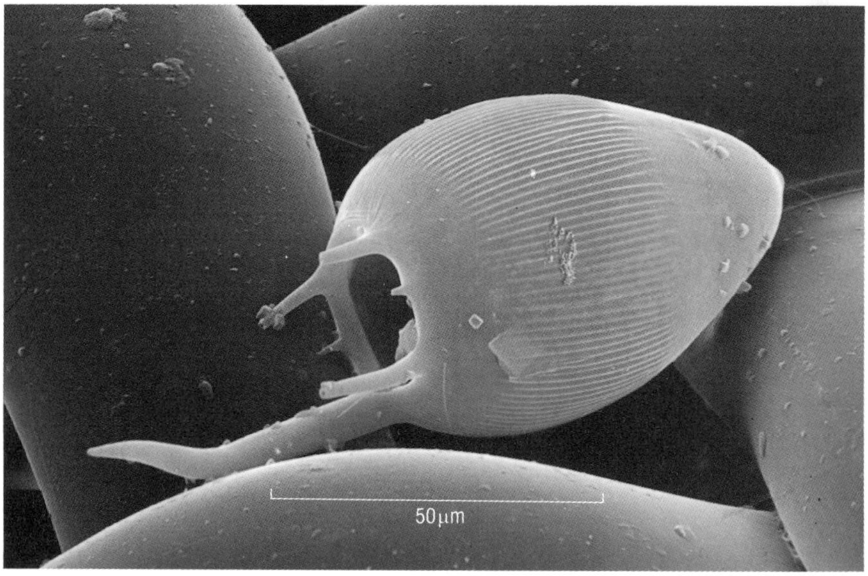

(b)

Figure 6.2 (a) Diatoms (phytoplankton skeletons, $SiO_2.nH_2O$). Common in deep-sea deposits under areas of high surface biological productivity.

(b) Radiolaria (zooplankton skeleton, $SiO_2.nH_2O$). Plentiful in deep-sea sediments (hence the name Radiolarian ooze), commonly accompanied by diatoms. The background is part of the sampling net.

The third nutrient, silica (or silicate), is used to build the skeletons of planktonic plants (diatoms) and animals (radiolarians) (Figure 6.2). The silica secreted by organisms is an amorphous form and it is hydrated, so its formula is commonly written as $SiO_2.nH_2O$ (and it is sometimes called opaline silica or opal). But for brevity we shall use SiO_2 for both solid and dissolved silica. As the organisms die or are consumed, the skeletal debris sinks through the water column and slowly dissolves, giving concentration–depth profiles like that in Figure 6.1 (c)

QUESTION 6.3 Why are well-stratified surface waters likely to be more rapidly depleted in nutrients than a well-mixed column? In which situation is there a well-developed pycnocline?

Many marine plants and animals form skeletons of calcium carbonate (Figure 6.3), so carbon is used for both the soft and hard parts of organisms. The biological utilization of carbon and calcium in the marine environment is a major component in the global cycles of these two elements. Both are so abundant in the seawater solution, however (Table 6.1), that they never limit primary production: the amounts used by organisms are small in relation to the total abundance, and the effects on the C:*S* and Ca:*S* ratios are not easy to detect (Section 3.1). The dissolved species of elements such as carbon and calcium are sometimes called **bio-intermediate constituents** because although they show some depletion in surface waters, they cannot be exhausted, even in regions of very high biological production.

Constituents whose concentrations in solution are unaffected by biological activity are sometimes called **bio-unlimited constituents**—and these are also the constituents that behave conservatively in seawater, such as sodium and chloride.

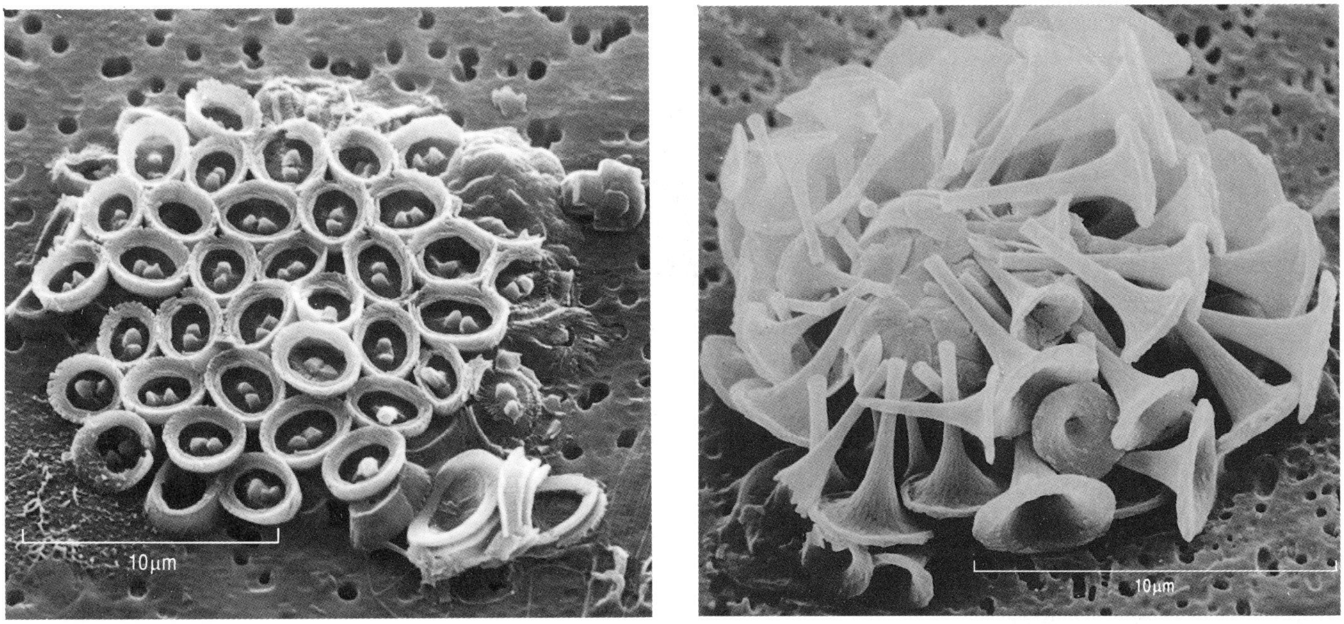

(a)

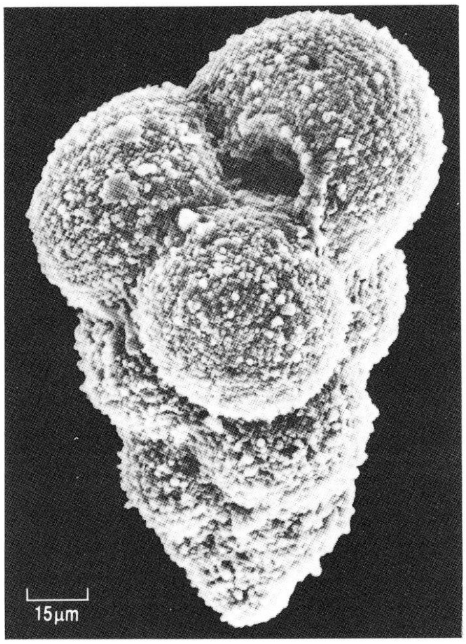

(b)

(c)

Figure 6.3 (a) Coccolithophores (phytoplankton skeletons, $CaCO_3$, *Discosphaera* sp. and *Coronosphaera* sp.). The coccosphere disaggregates during deposition and is rarely preserved intact. The component platelets are common in deep-sea sediments and in chalk deposits (e.g. the white cliffs of Dover).

(b) Foraminifera (zooplankton skeleton, $CaCO_3$ *Guembelitria* sp.). Foraminifera are another common constituent of deep-sea sediments.

(c) Pteropod (zooplankton skeleton, $CaCO_3$, aragonite—see later text). These are molluscs (gastropods), easily recognizable in sediments because of their relatively large size.

6.1.3 DISSOLVED GASES

Three-quarters of the mass of the atmosphere is concentrated in the lowest 10km, and this part of the atmosphere shows no variations in the proportion of its major constituents: nitrogen (78%), oxygen (21%) and argon (1%). (Concentrations of atmospheric constituents are given by volume.) Carbon dioxide is fairly uniformly distributed in the lower atmosphere but accounts for only about 0.03% of the total.

Figure 6.4 predicts the solubilities of the four most abundant gases in seawater at 24°C. An important feature of Figure 6.4 (a) is that the vertical scale is given in **partial pressure**, which is identical with percentage composition by volume: for example, if you took away all the gases except oxygen, the 21% oxygen would give a pressure of 0.21 atmospheres.

QUESTION 6.4 (a) What is the ratio of nitrogen to oxygen (i) in the atmosphere and (ii) in seawater? How much more or less soluble is nitrogen than oxygen?

(b) On the same reasoning, how much more or less soluble is argon than carbon dioxide?

Important: Note that concentrations of gases are given in $mg\,l^{-1}$ (p.p.m. by weight) in Table 6.1 and these are numerically not very different from the volumetric concentrations ($ml\,l^{-1}$) for oxygen, nitrogen and argon in Figure 6.4. That is because of the low densities of these gases: respectively 1.43, 1.23 and $0.77\,kg\,m^{-3}$ (so, to a first approximation, $1\,m^3$ weighs 1000g, 1 litre weighs 1 gram, and 1ml weighs 1mg).

Figure 6.4 shows that the solubility of CO_2 in seawater is many times greater than that of nitrogen and oxygen. The difference is due to the reactivity of CO_2 in seawater leading to the various carbonate and bicarbonate equilibria:

$$CO_2\,(gas) + H_2O \rightleftharpoons H_2CO_3 \rightleftharpoons H^+ + HCO_3^- \rightleftharpoons 2H^+ + CO_3^{2-} \qquad (6.2)$$

The information in Figure 6.4 is a useful starting point for the discussion of dissolved gases, but it must be treated with caution, as the data apply to seawater at one temperature only. The solubility of gases generally decreases with increasing temperature and salinity, and increases with increasing pressure. In addition, Figure 6.4 is based upon the assumption that there is equilibrium between atmosphere and ocean across the air–sea interface. At equilibrium, rates of gaseous diffusion are the same in both directions (i.e. there is no net flux of gas into or out of seawater) because the number of molecules of gas entering the seawater solution is equalled by the number of molecules escaping back to the atmosphere. This is probably valid to a first approximation for the four most abundant gases, but not for many that occur in much lower concentrations. These include the 'noble' gases, helium (He), neon (Ne), krypton (Kr) and xenon (Xe), as well as some others that are discussed later.

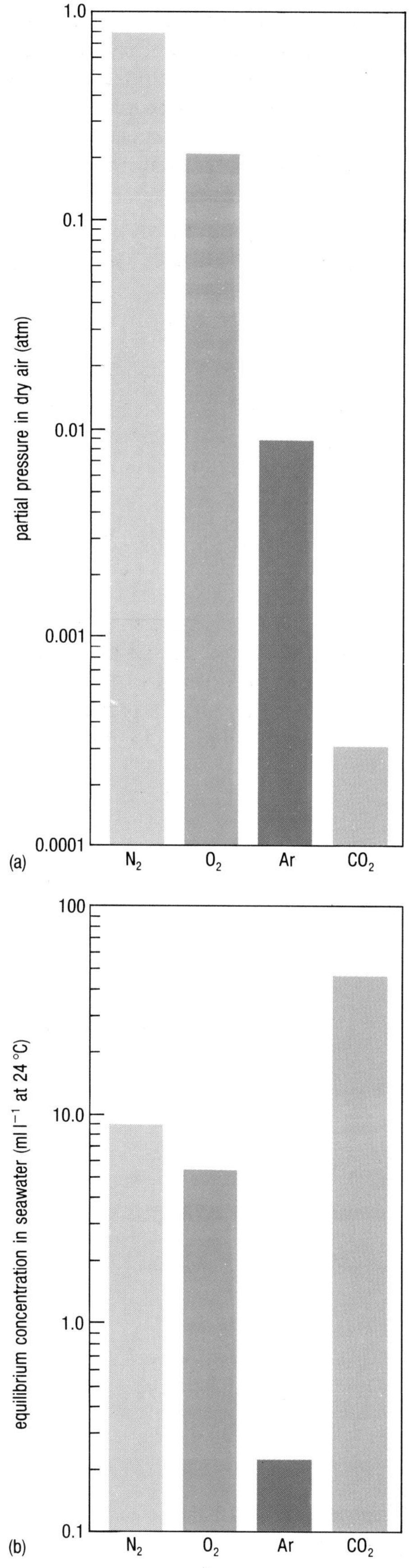

Figure 6.4 (a) Partial pressure (= volume proportions) of the four most abundant gases in the atmosphere, totalling together 99.93% of the atmosphere, the rest being made up of other minor gases. (b) Equilibrium concentrations by volume of these four gases dissolved in seawater at 24 °C as controlled by their atmospheric partial pressures. Note the different scales on the vertical axes.

The distribution of gases at deeper levels in the oceans is achieved mainly by currents and by turbulent mixing rather than by diffusion. Downward redistribution is slow, however, and the oceans may take hundreds or thousands of years to equilibrate fully with the ambient atmosphere—in other words, it takes a long time for the effects of changes in the processes that control gas exchange at the air–sea interface to be 'transmitted' throughout the oceans.

Biological activity plays an important role in the redistribution of oxygen and carbon dioxide below the surface, and largely determines the form of their concentration–depth profiles.

Oxygen

Ocean surface waters are consistently supersaturated with oxygen (Figure 6.5), partly due to liberation of oxygen during photosynthesis, but mainly as a result of air bubbles formed at the crests of waves being forced down into the water column, where part of the gas they contain is driven into solution by the increased hydrostatic pressure.

Near the bottom of the photic zone, there is a balance between the amount of carbon that phytoplankton fix by photosynthesis and the amount they dissipate in respiration. The depth at which this balance occurs is called the **compensation depth**, and it can also be defined as the depth at which the amount of oxygen produced by phytoplankton during photosynthesis equals the amount they consume in respiration over a 24-hour period. In short, for the phytoplankton population in a particular location, reaction 6.3 (p. 96) has reached a balance: for every mole of

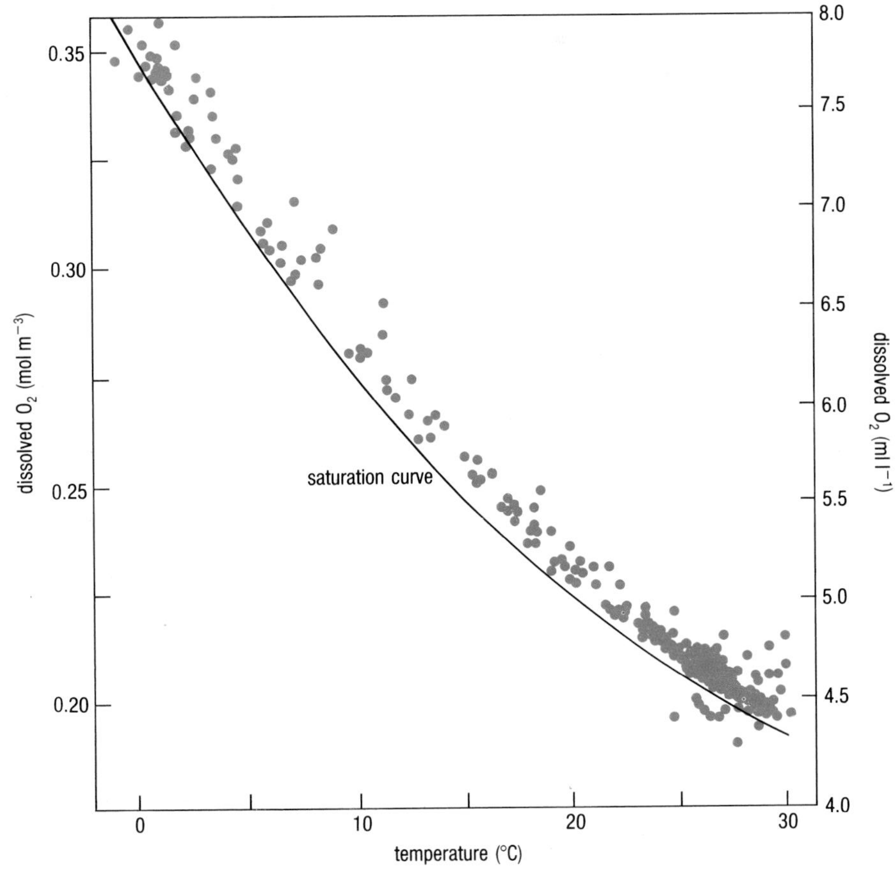

Figure 6.5 The saturation curve for oxygen (solid black line) and measured concentrations (blue dots) in ocean surface waters, determined as part of the GEOSECS programme.

oxygen produced (or mole of carbon fixed) in photosynthesis, another mole is used (or dissipated) in respiration:

$$CO_2 + H_2O \underset{\text{metabolic energy (respiration)}}{\overset{\text{light energy (photosynthesis)}}{\rightleftharpoons}} \underset{\text{organic matter}}{(CH_2O)_n} + O_2 \quad (6.3)$$

Photosynthesis does not cease at the compensation depth, but below it there can be no net phytoplankton growth, because by definition more oxygen is being used in plant respiration than is produced by photosynthesis. Algae can survive to considerable depths—viable plant cells have been recovered from depths of several thousand metres—but they cannot actually grow once they sink below the compensation depth. For practical purposes, then, the compensation depth can be taken to represent the lower limit of the photic zone.

At greater depths, oxygen continues to be consumed in the respiration of both animals and plants and in the microbial decomposition (oxidation) of organic detritus; and it is not being replenished, because photosynthesis declines to negligible levels below the photic zone. An **oxygen minimum layer** develops where maximum abstraction of oxygen has occurred, generally between 500 and 1000m depth. In some areas, such as the northern Indian Ocean and the eastern tropical Pacific (Figure 6.6), the water at these depths is very oxygen-deficient, and in extreme cases it can become completely anoxic. At greater depths in the open oceans, oxygen levels rise again, because of the input of cold, dense oxygenated water

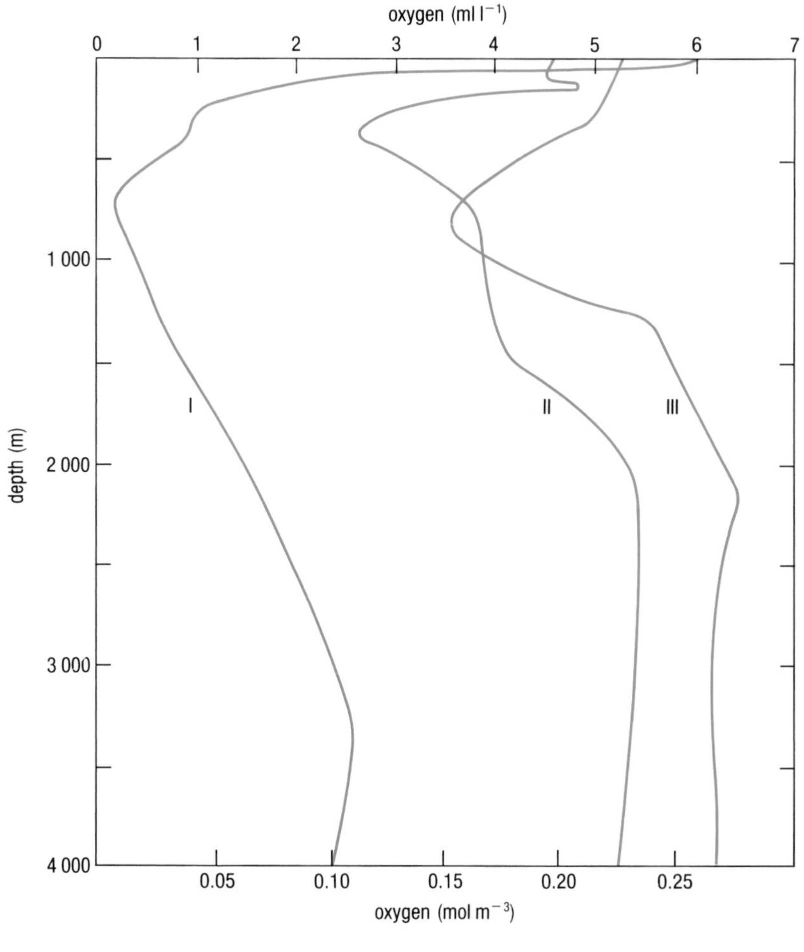

Figure 6.6 The vertical distribution of dissolved oxygen (concentration in ml l^{-1} and mol m^{-3} (i.e. 10^{-3} mol l^{-1}) I: South of California; II: eastern part of the South Atlantic; III: Gulf Stream.

sinking in polar regions (*cf*. Figure 2.13). The vertical distribution of oxygen varies considerably from place to place (Figure 6.6), but in general it is almost a mirror image of those for phosphate and nitrate.

Where strong oxygen minimum layers coincide with the sea-floor, anoxic sediments may be deposited. This typically occurs along continental margins, where there is high biological productivity. In addition, where basins are isolated by shallow barriers (sills) from oxygenated deep-sea circulation (e.g. the Black Sea), anoxic sediments form at all depths below the level at which oxygen is exhausted.

QUESTION 6.5 (a) Why is water sinking in polar regions more highly oxygenated than that elsewhere?

(b) On what grounds might we infer a greater degree of biological activity to have occurred in the water column represented by profile I in Figure 6.6 than in the other two?

Carbon dioxide

Just as in the case of oxygen, the lower the temperature, the more CO_2 goes into solution. Below the thermocline, however, where temperature is virtually constant, the solubility of carbon dioxide becomes almost entirely a function of pressure: increased pressure forces more CO_2 into solution to form carbonic acid and its dissociation products (reaction 6.2). This example of Le Chatelier's principle is well known to consumers of bottled or canned beverages that froth or fizz upon being opened, as the CO_2 gas escapes when the pressure is released.

Seawater in equilibrium with atmospheric CO_2 is slightly alkaline, with a pH of around 8.1–8.3. The pH may rise slightly through the rapid abstraction of CO_2 from surface waters during photosynthesis (reaction 6.3), but it does not normally exceed 8.4 except in tidal pools, lagoons and estuaries. We have seen that below the photic zone, the CO_2 absorbed in photosynthesis is exceeded by the CO_2 released in respiration. As CO_2 concentrations increase, so pH falls, typically to values of about 7.8–8.0. It can reach values of 7.5 or less in waters of reduced salinity or in anaerobic conditions, where bacteria using reduction of sulphate as a source of oxygen for the decomposition of organic matter release H_2S into solution (*cf*. Section 3.1.1). However, when the sulphate has been used up, decomposition of organic matter under anaerobic (anoxic) conditions involves the reduction of CO_2 itself, and leads to the formation of hydrocarbons, such as methane CH_4. Under these conditions, the pH may rise to values as high as 12. We look at pH in more detail in Section 6.3.2.

Some minor gases

All but one of the gases listed in Table 6.2 (overleaf) are produced by organisms in surface waters in which the gases are oversaturated with respect to their atmospheric concentrations. So, for equilibrium to be maintained, they must escape to the atmosphere; in other words, their net flux is from sea to air.

The exception in Table 6.2 is *sulphur dioxide*, whose net flux is from air to sea. Its sources include volcanism and industry (fossil fuel burning and metal smelting), and oxidation of natural organic sulphur compounds. In the atmosphere, it is oxidized to SO_3, which quickly combines with water to form sulphuric acid aerosols which contribute to the problem of acid

Table 6.2 World-wide sea–air fluxes for some gases.

Gas	Total oceanic flux ($g\,yr^{-1}$)	Direction of net flux
SO_2	1.5×10^{14}	air→sea
N_2O	1.2×10^{14}	sea→air
CO	4.3×10^{13}	sea→air
CH_4	3.2×10^{12}	sea→air
CH_3I	2.7×10^{11}	sea→air
$(CH_3)_2S$	4.0×10^{13}	sea→air

rain; so, much of its input to the oceans may be as sulphate ions (SO_4^{2-}) as well as gaseous SO_2.

Surface seawater is generally oversaturated with *nitrous oxide* (N_2O), because of bacterial activity, and the resulting flux from sea to air may be important in the oceanic nitrogen budget. The rate at which fixed nitrogen enters the oceans from river inflow and rain is about $8\times10^{13}\,g\,N\,yr^{-1}$. Approximately 10% of this ($9\times10^{12}\,g\,N\,yr^{-1}$) is removed to marine sediments as organic nitrogen compounds in undecomposed organic detritus, leaving 90% still unaccounted for.

QUESTION 6.6 What is the sea→air flux of N_2O given in Table 6.2, in terms of $g\,N\,yr^{-1}$? Use relative atomic masses N = 14, O = 16. Does this make up the balance of the nitrogen input to the oceans that is not removed to the sediments?

Carbon monoxide (CO) and *methane* (CH_4) provide an interesting contrast. Their concentrations in surface seawater are similar, but the atmospheric concentration of carbon monoxide is much *less* than that of methane. The concentration gradient across the interface is therefore much greater for CO than for CH_4, which accounts for the order of magnitude difference in their fluxes (Table 6.2). The sea→air flux of these gases is not an important component of the global carbon budget—the flux of CO in Table 6.2 represents only about one-fifth of the annual input of carbon monoxide to the atmosphere from human activity.

Methyl iodide (CH_3I) and *dimethyl sulphide* ($(CH_3)_2S$) are unstable in oxygenated environments. They are produced by organic activity near the sea-surface and persist long enough to pass into the atmosphere, where they are decomposed by ultraviolet radiation.

6.1.4 DISSOLVED GASES AS TRACERS

Whatever the present position of a subsurface body of water, it must at some time have been at the surface, where diffusion across the air–sea interface will have determined its dissolved gas content. Once the water has sunk away from the surface and become isolated from the atmosphere, the concentrations of dissolved gases will change as a result partly of mixing and partly of biological or other reactions.

Oxygen is used as a tracer because it is abundant, biologically important and easily measured. The longer a water mass is isolated from the atmosphere, the lower its oxygen content becomes. By tracking back along the concentration gradient of oxygen, the source region of the water mass can be located, and the changes that have gone on within the water mass since its isolation from the surface can be inferred.

QUESTION 6.7 (a) Why is oxygen a non-conservative dissolved constituent, and why does its concentration progressively decrease with time after the water in which it is dissolved has left the surface?

(b) Use the distribution of dissolved oxygen content shown on Figure 6.7(a) to identify some of the main deep water masses, in particular North Atlantic Deep Water, Antarctic Intermediate Water, and Antarctic Bottom Water (see also Figures A1, 2.6, and 3.3).

(c) Does the distribution of dissolved oxygen content in the North Pacific (as shown in Figure 6.7(b)) suggest the presence of a source region of deep water there?

Figure 6.7 Sections showing dissolved oxygen (ml l^{-1}) in (a) the western Atlantic (*cf.* Figures 2.6 and 3.3) and (b) the Pacific (about 170° W).

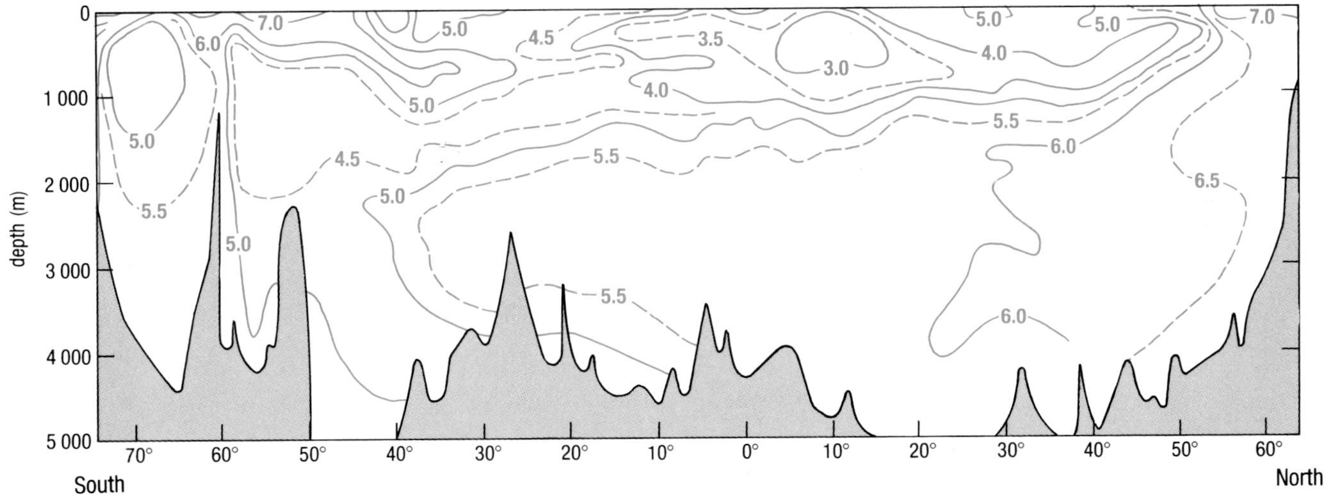

(a)

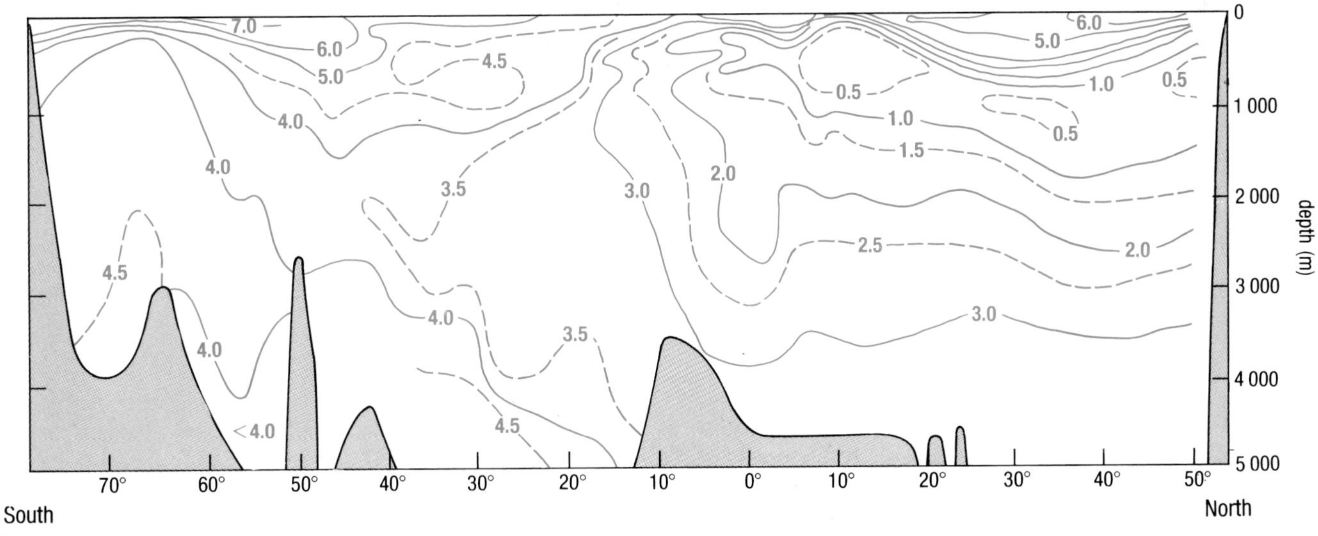

(b)

6.2 SOURCES AND SINKS, OR WHY THE SEA IS SALT

In this and subsequent Sections we shall look at the ways in which supply and removal of dissolved constituents contribute to the overall compositional balance of seawater, and then look at some of the chemical reactions in which the different constituents participate.

If rivers are the chief source of the dissolved salts in seawater, why is seawater not simply a concentrated version of the average composition of all rivers?

The most important part of the answer lies in the chemical behaviour of water as it moves through the hydrological cycle. It picks up salts in solution during weathering on land, and precipitates those salts, in different amounts and at different rates, after it reaches the sea again.

6.2.1 COMPARISON OF SEAWATER WITH OTHER NATURAL WATERS

Figure 6.8 shows the average concentrations of the principal dissolved constituents in rainwater, river water and seawater. The averages of rainwater and river water conceal considerable variations, but the basic pattern is the same all over the world.

QUESTION 6.8 (a) How many times more dilute than seawater is (i) rainwater (ii) river water?

(b) Is rainwater or river water closer to seawater in composition?

Thus to change rainwater into river water clearly requires the addition of substantial amounts of certain constituents, and these are provided mainly by the chemical weathering of rocks. Rainwater contains dissolved gases, particularly CO_2 and SO_2, both of which form acidic solutions in water, so that rainwater is a weak acid (pH≈5.7). When rain falls on the land, the acidity is neutralized by reaction with minerals in soils and rocks:

$$\underset{\text{(calcite, a common mineral in sedimentary rocks)}}{CaCO_3} + \underset{\text{(from rainwater)}}{CO_2 + H_2O} \rightarrow \underset{\text{(in solution)}}{Ca^{2+} + 2HCO_3^-} \qquad (6.4)$$

$$\underset{\text{(albite, a common mineral in igneous and metamorphic rocks)}}{2NaAlSi_3O_8} + \underset{\text{(from rainwater)}}{2CO_2 + 3H_2O} \rightarrow \underset{\text{(kaolinite, a clay mineral)}}{Al_2Si_2O_5(OH)_4} + \underset{\text{(in solution)}}{2Na^+ + 2HCO_3^-} + \underset{\text{(silica, partly in solution)}}{4SiO_2} \qquad (6.5)$$

The two representative examples shown in reactions 6.4 and 6.5 simplify the real situation, but they account broadly for the processes by which rainwater is transformed into river water. The exceptionally large increases in the concentrations of Ca^{2+} and HCO_3^- between rainwater and river water (Figure 6.8) arise from the fact that these ions can be produced from weathering both of carbonates (reaction 6.4), and of calcium-bearing silicates (in reactions analogous to 6.5).

6.2.2 SEAWATER AND RIVER WATER

Seawater contains about 300 times more dissolved salts than average river water (Question 6.8), and a glance at Figure 6.8 shows the mix of elements dissolved in river water to be very different from that in

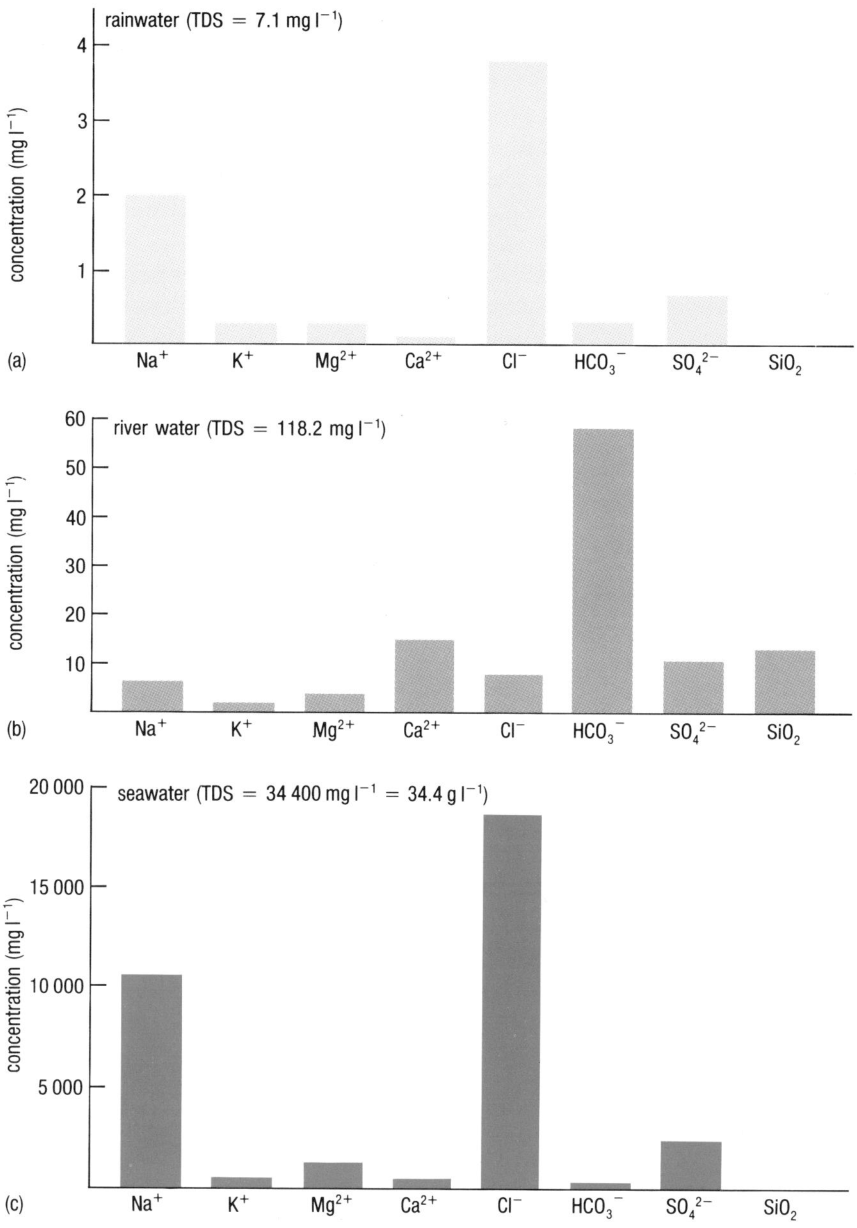

Figure 6.8 The average chemical composition of seawater, river water and rainwater for up to eight dissolved constituents, some at concentrations too low to appear (TDS = total dissolved salts). Note that total concentrations increase from rainwater to river water to seawater. Arrows indicate the magnitude of the change of scale from top to bottom.

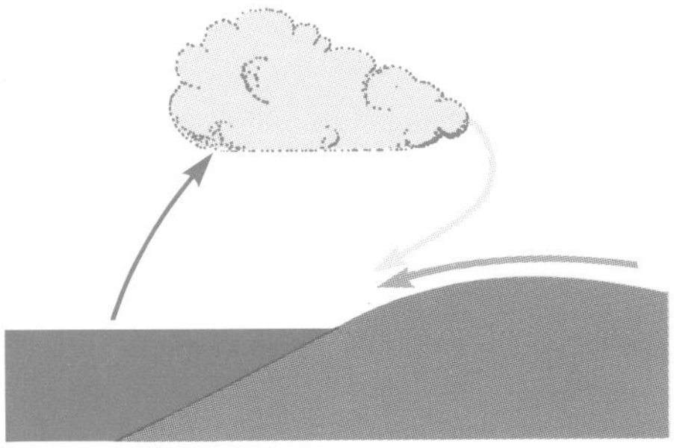

Figure 6.9 Of the dissolved constituents entering the oceans from rivers, some are derived from rock weathering and some are recycled marine salts.

seawater. In the marine environment, substantial amounts of HCO_3^-, Ca^{2+} and SiO_2 in particular must be removed from solution. We have established that some of the salts in river water come from chemical weathering of surface rocks, the remainder being recycled sea salts contributed by rainwater (Figure 6.9). We shall now try to quantify the relative contributions from these two sources.

The average chloride content of continental crustal rocks is in the order of 0.01%, and so only a very minute proportion of the chloride in river water comes from weathering. It follows that virtually all the chloride in river water (Figure 6.8) must be from sea salt recycled via oceanic aerosols (Section 2.2.1). This enables us to correct river water compositions for recycled or **cyclic salts**.

How might we do this?

We can do it by applying the constancy of composition of seawater for major constituents (Section 3.1). Our basic assumptions are that all the chloride in river water is recycled from the oceans by rain (and snow), and that the other constituents are recycled in the same proportions as they occur in seawater. These assumptions have been applied in Figure 6.10 to 'correct' the measured concentrations in river water by subtracting the contributions from cyclic salts. What remains is the contribution from weathering.

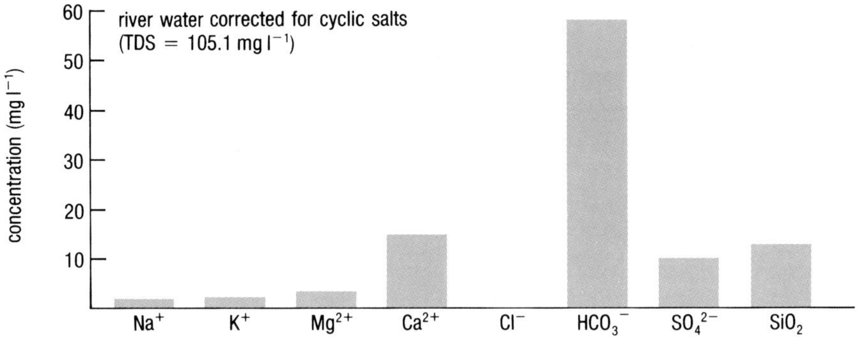

Figure 6.10 The average chemical composition of river water after correction for cyclic salts (*cf.* middle histogram of Figure 6.8).

QUESTION 6.9 When the composition of river water is corrected for cyclic salts, as in Figure 6.10, the proportions of the major ions differ from those in seawater by an even greater amount than shown in Figure 6.8. Which of the following relative cation and anion abundance patterns apply to river water (corrected for cyclic salts) and which to seawater?

(i) $Na^+ > Mg^{2+} > Ca^{2+}$

(ii) $Ca^{2+} > Mg^{2+} > Na^+$

(iii) $HCO_3^- > SO_4^{2-} > Cl^-$

(iv) $Cl^- > SO_4^{2-} > HCO_3^-$

6.2.3 ORIGIN OF THE CHLORIDE

It is easy enough to see how the major ions—sodium, potassium, magnesium and calcium—in the oceans can be derived from weathering of rocks, because these elements have a high abundance in the Earth's crust (Table 3.2). By contrast, only a negligible proportion of the chloride in river water comes from weathering, which is why we set chloride to zero when correcting for cyclic salts.

So where does the chloride come from?

The answer lies in volcanism. HCl is an important constituent of volcanic gases. Early in the Earth's history, volcanism was more widespread than it is now, because the Earth as a whole was hotter. Large quantities of this very soluble gas were emitted and quickly dissolved in the oceans. Chloride is classified as an **excess volatile**, a constituent of seawater that cannot be accounted for by rock weathering.

6.2.4 THE SODIUM BALANCE

Are there more constituents in seawater for which a source other than weathering should be sought, and if so, how can we tell?

A means of testing for an additional source of an element in seawater is to make a **mass balance calculation**. The total amount of the element being added to the oceans by continental weathering is compared with the amount of seawater. If there is more of the element in seawater than can be accounted for by rock weathering, then there must be an additional source of that element. If the amount in seawater is the same or less, it is not necessary to seek a source additional to rock weathering, though such a source may well also exist.

One of the simplest approaches is to work out the **sodium balance**. We assume that there is no source other than rivers for sodium in seawater, and calculate the amount of continental crustal rock that has to be weathered to provide this sodium. We can then compare the results for other elements with that for sodium. The calculation involves a number of simplifying assumptions, but this does not matter because we wish merely to identify those elements in seawater for which there is likely to be an additional source.

The first step is to estimate how much crustal rock must be weathered to provide the sodium in 1 litre of seawater. Tables 3.1 and 6.1 give an approximate average of 11g of sodium per litre of seawater, which is accurate enough for our simple first order sums. The average concentration of sodium in crustal rocks is 2.4% (Table 3.2), and we can assume that to be representative of continental crust. So, there are 2.4g of sodium in 100g of average (continental) crustal rock.

It is estimated that approximately three-quarters of the sodium in rocks goes into solution on weathering and ultimately reaches the sea. The rest remains chemically combined in the minerals of detrital (fragmental) sediments.

QUESTION 6.10 75% of 2.4 is 1.8. So from every 100g of weathered rock, about 1.8g of sodium goes into solution. To the nearest 100g, how many grams of rock must be weathered to provide the 11g of sodium in 1 litre of seawater?

The next step is to see whether the amount of average crustal rock that provides the sodium in a litre of seawater can also provide the other dissolved constituents in that same litre. This has been done for selected elements in Table 6.3 (overleaf). Look at the last column of Table 6.3. The 'percentage in solution' for sodium is close to 75, because that provided the basis of the calculations leading to the answer you should have obtained in answering Question 6.10.

Now examine the 'percentage in solution' figures for the four elements below sodium in Table 6.3. Do they suggest that continental weathering is an adequate source of supply for these elements in seawater?

According to Table 6.3, less than 10% of any of those elements in average crustal rock is required to go into solution on weathering, in order to account for their concentrations in seawater. Perhaps you can see why the approximations used in the sodium balance calculation do not matter greatly: you could increase the 'percentage in solution' figure for sodium to 100, or reduce it to 50 or even less, and it would not affect the conclusion we have just reached.

Table 6.3 Comparison of elements in seawater and continental crustal rock.

Element	In continental crust		In seawater (gl^{-1})	Percentage in solution*
	%	g in 600g rock		
Na	2.4	14.4	10.76	74.7
K	2.1	12.6	0.387	3.1
Ca	4.1	24.6	0.413	1.7
Mg	2.3	13.8	1.294	9.4
Sr	0.038	0.23	0.008	3.5
Se	5×10^{-6}	3×10^{-5}	$\sim10^{-7}$	0.3
As	2×10^{-4}	1.2×10^{-3}	$\sim10^{-6}$	0.08
Pb	1.25×10^{-3}	7.5×10^{-3}	$\sim10^{-8}$	0.0001
Zn	7×10^{-3}	4.2×10^{-2}	$\sim10^{-7}$	0.0002
Cu	5.5×10^{-3}	3.3×10^{-2}	$\sim10^{-7}$	0.0003
Co	2.5×10^{-3}	1.5×10^{-2}	$\sim10^{-8}$	0.00007
Cl	0.013	0.078	19.353	24800
S	0.026	0.156	0.885	567
Br	0.00025	0.0015	0.067	4470
B	0.0003	0.0018	0.0046	256

*Percentage in solution $= 100\times\frac{\text{g per litre seawater}}{\text{g per 600g rock}}$

The elements in the first group of Table 6.3 are major constituents in seawater. Elements in the second group are minor and trace constituents, and their 'percentage in solution' values are very small—well below 1. Indeed, at first sight it is surprising that for the first two groups in Table 6.3 the figures in the last column are mostly so small. They seem to imply that for many elements only a minute proportion is dissolved on weathering.

A more likely explanation is that dissolved constituents are removed at different rates from seawater: the lower the 'percentage in solution', the more efficient the inorganic or biological removal processes are likely to be—and the shorter the residence time of a particular constituent in the ocean. This is the only way of reconciling the information for river water and seawater in Figures 6.8 and 6.10. It is quite obvious that seawater is not simply a more concentrated form of river water. If it were, then HCO_3^- rather than Cl^- would be the principal anion, and Ca^{2+} rather than Na^+ the principal cation. It follows that the residence times of calcium and carbon in the oceans are much shorter than those of sodium and chloride (see Section 6.2.5).

The last group in Table 6.3 are all anion-forming elements in seawater. Headed by chloride, these obviously represent the excess volatiles, those constituents whose concentrations in seawater cannot be accounted for by rock weathering alone. SO_2, HBr and volatile boron compounds are all known to be components of volcanic gases, along with CO_2, nitrogen, argon, hydrogen, and of course H_2O. Water in present-day volcanic gases is mainly recycled from the atmosphere and hydrosphere, much of it via subduction zones at destructive plate margins; but early in the Earth's history, most of the H_2O in volcanic gases must have originated from the planet's interior. In other words, the world's water inventory (Figure 1.3) is the result of planetary de-gassing, as outlined at the start of Chapter 1.

More sophisticated mass balance calculations take into account factors such as the composition of volcanic gases and the composition and rate of deposition of marine sediments. They have been used to identify additional excess volatiles among the minor and trace constituents of seawater, such as selenium, arsenic and lead, all of which have very low 'percentage in solution' figures in Table 6.3 (remember the sodium balance is a very simple model and these low values do not preclude the existence of a source additional to weathering for particular dissolved constituents). Another interesting case is that of manganese: deep-sea sediments contain greater concentrations of manganese than continental rocks, and more manganese is being deposited than is supplied by weathering.

Can you suggest another source for manganese (and other elements) in the oceans, one that has only been discovered comparatively recently?

Hydrothermal activity at ocean ridge crests and other sites of oceanic volcanism is known to supply some elements to seawater, notably calcium and manganese; it also provides a sink for some elements, notably magnesium, large amounts of which are removed from solution by this mechanism.

6.2.5 CHEMICAL FLUXES AND RESIDENCE TIMES

Implicit in the arguments set forth in preceding Sections is the assumption that the oceans are chemically in a long-term **steady state**. This means that the rate of addition of dissolved constituents to seawater is balanced by their rate of removal, so that concentrations do not change significantly with time. (A useful analogy is with a bottling machine: empty bottles go in at one end, full bottles come out at the other, but the number on the conveyor belt is always the same, no matter how fast the machine is moving, and photographs of it taken at different times would all look alike.) There is evidence that a chemical steady state condition has characterized the oceans since early in the Earth's history, and that the composition of seawater has not varied significantly over the past several hundred million years.

The concept of residence time was introduced in Section 1.2, in relation to the hydrological cycle. How can it be applied to dissolved constituents in the oceans?

If the oceans are in a steady state, then the rates of supply and removal of dissolved constituents must be equal. The residence time of a dissolved constituent is given by:

$$\frac{\text{total mass dissolved in the oceans}}{\text{rate of supply (or removal)}}$$

and as rates of supply or removal are usually annual rates, residence times are normally given in years.

A traditional starting point for residence time calculations has been that rivers are the only source of supply of dissolved constituents. This basic assumption is still valid enough for most dissolved constituents to provide reasonable estimates of residence times, even though it is now known that hydrothermal solutions provide significant amounts of some elements.

The residence time of an element in seawater can therefore be estimated by dividing its mass in the oceans by its annual input from rivers. The annual flux of each element into the oceans via rivers can readily be calculated from the total annual inflow of river water to the oceans (e.g. Figure 1.3), multiplied by the average concentration of that element in river water (e.g. Figure 6.10). The total mass of each element in the oceans is also easily calculated from data for average concentrations in seawater and the total mass of water in the oceans (*cf.* Table 6.1). Oceanic residence times for several elements are presented in Table 6.4.

Table 6.4 River fluxes and residence times of some dissolved constituents in seawater.

Constituent	River flux* ($\times 10^8$ t yr^{-1})	Mass in ocean** ($\times 10^{14}$ t)	Residence time ($\times 10^6$ yr) (uncorrected)*	(corrected)
Na^+	2.05	144	70.2	210
K^+	0.75	5	6.7	10
Ca^{2+}	4.88	6	1.23	1
Mg^{2+}	1.33	19	14.3	22
Cl^-	2.54	261	103	(∞)
HCO_3^-	18.95	1.9	0.1	0.1
SO_4^{2-}	3.64	37	10.2	11
SiO_2	4.26	0.08	0.02	
Fe	0.22	0.000014	0.00006	
Mn	0.001	0.00002	0.0002	
Cu	0.0007	0.000021	0.03	
Co	0.001	0.000001	0.0001	
Zn	0.0007	0.000042	0.006	

*These values are not corrected for cyclic salts.

**Amounts differ somewhat from those in Table 6.1 (see accompanying text to that Table).

QUESTION 6.11 Major constituents in Table 6.4 have two residence times: on the left are values derived using the river flux uncorrected for cyclic salts (i.e. using concentrations in river water such as those in Figure 6.8); while on the right are values obtained after correcting the river flux for cyclic salts (i.e. using concentrations in river water such as those in Figure 6.10).

(a) For how many constituents is the residence time significantly affected by this correction?

(b) It has been estimated that some 2.5×10^8 t of calcium are added to the oceans annually by hydrothermal circulation. How would that affect the residence times in Table 6.4?

Residence times in Table 6.4 are approximate only. The averages used conceal wide variations and in many cases are based on limited data. Moreover, the basic assumption is not wholly valid: rivers are not the sole source of dissolved constituents in seawater. The real residence time of chloride is neither infinite nor about 100 million years (Table 6.4), but somewhere in between. There is evidence that chloride is removed from seawater during hydrothermal circulation and, of course, it is also removed when **evaporite** salts are deposited. These losses are balanced by the continued emission of HCl from volcanoes, which adds chloride to the atmosphere–ocean system.

Figure 6.11 shows that there is a broad correlation between residence time and concentration in seawater. Most of the major constituents (for which the 'percentage in solution' is high—Table 6.3), have long residence times and remain in seawater for the order of 10^6–10^8 years, whereas minor and trace constituents have short residence times and are removed from seawater in 10^3–10^4 years or less—and their 'percentage in solution' is low.

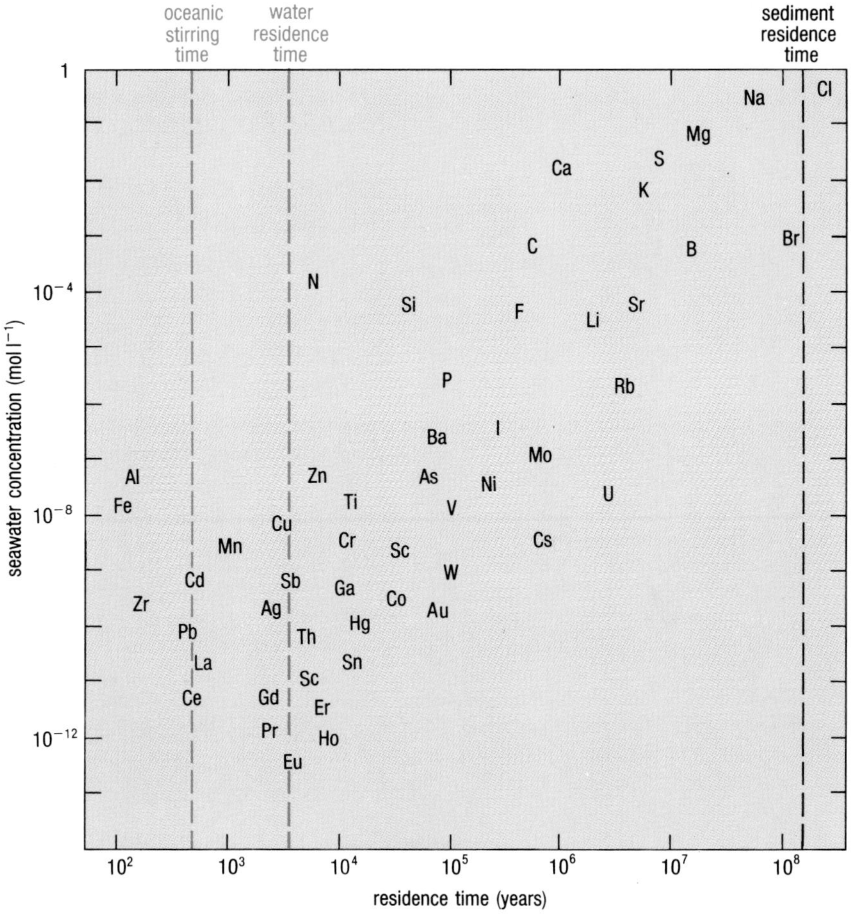

Figure 6.11 Graph showing the broad correlation between concentration (mol l-1) and residence time for several elements in seawater. For water residence time, see Question 1.3 (b) and its answer; for oceanic stirring time and sediment residence time, see the following text.

The oceanic stirring time for water in the oceans is shown as about 500 years in Figure 6.11. This represents the average time water spends in the deep oceans, before it returns to the surface. Carbon-14 dating of dissolved carbon species is used to work out these mixing or turnover times, as they are also known. Earlier estimates gave values approaching 2000 years, but more recent measurements and calculations suggest much shorter periods: about 300 years for the deep Atlantic, about 600 years for the deep Pacific, and the average for the oceans as a whole is about 500 years.

For individual dissolved constituents, the residence time is the average length of time spent in the seawater solution. Dissolved constituents are added from rivers and other *sources*. They reside in solution for a time before being removed to the sediments and rocks of the sea-floor, which are the *sinks*. The reactions which remove dissolved constituents to the sinks are often called reactions of **reverse weathering**, because they have the reverse effect to reactions which supply elements to seawater as a result of weathering on land (e.g. reactions 6.4 and 6.5). Removal mechanisms include inorganic precipitation and reactions between dissolved material and solid particles, as well as biological processes. Marine organisms can concentrate minor and trace elements to very high levels in their soft tissues (see Section 6.3.4) and thus contribute to their removal from seawater; however, removal will be only temporary if the organism decomposes in the water column rather than being preserved in sediments. The residence time for sediments is longer than for most dissolved constituents, because it approximates to the lifetime of ocean basins. This is the length of time that a piece of sea-floor exists, on average, before being subducted at a destructive margin, and is generally between 100 and 200 million years.

QUESTION 6.12 (a) What is the distinction between the oceanic stirring time and the residence time for water in the oceans?

(b) Are there any dissolved constituents whose residence times are too short to allow them to be mixed throughout the oceans?

Dissolved constituents with short residence times are often said to be more reactive, or to have greater reactivity, in seawater than constituents with long residence times. In this context, reactivity refers simply to the ease with which a constituent enters insoluble phases and is precipitated, rather than to the chemical properties of the element concerned. For example, in its elemental state sodium is chemically a great deal more reactive than iron—but iron has a very much shorter residence time than sodium in the oceans and is therefore said to have greater reactivity in seawater.

6.3 CHEMICAL AND BIOLOGICAL REACTIONS IN SEAWATER

The form or **speciation** of dissolved constituents in seawater is very important in determining how they interact, and this in turn determines how long they remain in solution. Most of the dissolved constituents are in ionic form. The ions are kept apart because water has a high dielectric constant (Table 1.1), and each ion is surrounded by a sheath of water molecules called a **hydration sphere**, which has a diffuse outer boundary (Figure 6.12). The size of the hydration sphere depends on the radius and charge of the ion, which determine the charge per unit area, or charge density. So we can make some simple generalizations.

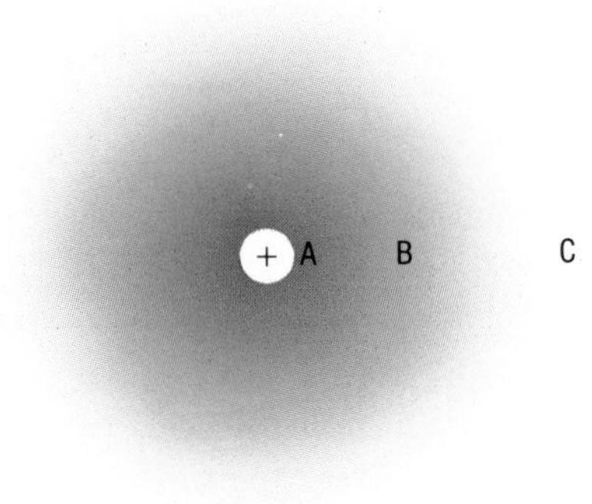

Figure 6.12 Pictorial representation of hydration of an ion. Water molecules in zone A are tightly held, in zone B less strongly bound, and in zone C hardly affected at all. There is probably a fairly well-defined boundary between zones A and B, but that between B and C is very diffuse.

1 **Anions** typically have lower charge densities than cations, because they are generally larger than the parent atom or molecule, having gained one or more electrons. **Cations** are generally smaller than the parent atom, because they have lost one or more electrons, and so cations have larger hydration spheres relative to their size than anions.

2 The greater the charge on an ion of given radius, the larger its hydration sphere relative to the size of the ion.

QUESTION 6.13 Which of the ions listed below is likely to have the largest hydration sphere relative to its size, and which the smallest?

Mg^{2+}	radius = 66 pm (pm = picometre = 10^{-12} m)
Na^{+}	radius = 97 pm
Cl^{-}	radius = 181 pm

6.3.1 INTERACTIONS BETWEEN DISSOLVED SPECIES

Cations and anions in solution experience mutual electrostatic attraction and repulsion because of their ionic charges. Such interactions are inversely proportional to the square of the distance separating the ions, and will be vanishingly small in very dilute solutions where the ions are widely separated. In solutions as saline as seawater, however, interactions between dissolved species cannot be ignored. It is these interactions which determine the speciation of dissolved constituents, and their overall effect is to decrease the availability of ions for chemical reactions, whether inorganic or biological.

Figure 6.13 summarizes in diagrammatic form the three main types of interaction possible between ions in solution.

1 For ions of strongly ionic salts (strong electrolytes) such as sodium chloride, the only interaction is that of mutual electrostatic attraction and repulsion between ions that otherwise behave as independent entities; and the hydration spheres remain intact.

2 Some species may form ion pairs, in which the hydration spheres of the constituent ions remain largely intact, and the result can be either a

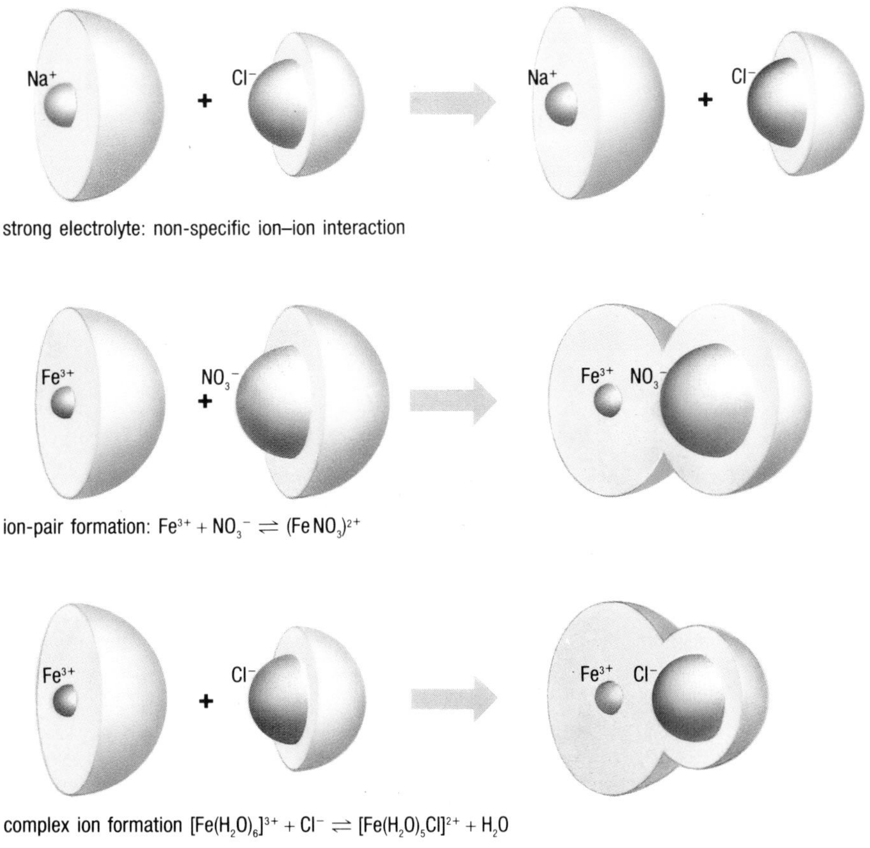

Figure 6.13 Diagrams illustrating different types of ionic interaction in seawater. Note that each ion is surrounded by a hydration sphere (*cf.* Figure 6.12). (a) General non-specific interaction. (b) Ion-pair formation. (c) Complex ion formation.

neutral or a charged species. It is uncommon for ion pairs to form between two monovalent ions; they are usually formed between two polyvalent ions or between one polyvalent and one monovalent ion.

3 Complex ion formation: although there is no particularly clear division between the formation of complexes and ion pairs, there are two principal differences. First, the bonding in a complex is covalent, rather than electrovalent as in an ion pair; and secondly, when a complex is formed in solution the hydration spheres of the two or more entities making up the complex merge to form a joint hydration envelope.

In summary, if part of the total amount of a dissolved ion is interacting with other solution species, then this fraction will not be free to participate in chemical reactions. The extent of these interactions and their effect on the chemical reactivity of dissolved species depend on the nature of the ions involved. In general, for solutions of a concentration equivalent to seawater, the fraction of the total concentration of an ionic species that is free to react decreases as its charge increases.

The 'effective' concentration, therefore, is almost invariably less than the true concentration. For unassociated ions (e.g. Na^+), subject only to electrostatic interaction, this effective concentration (known as the *activity* in quantitative treatments of solution equilibria) may be in the region of 70% of the true concentration, whereas for some complex-forming multivalent ions (e.g. Al^{3+}) it may be as low as 5–10%.

It has been known since the early 1960s that there is a considerable degree of ion pairing between several of the eight most abundant dissolved species in seawater: Na^+, K^+, Mg^{2+}, Ca^{2+}, Cl^-, HCO_3^-, CO_3^{2-}, and SO_4^{2-}, which between them make up over 99% of dissolved salts. Concentration data and equilibrium constants for interactions between these four cations and four anions have been used in fairly simple but lengthy calculations to arrive at the results summarized in Table 6.5.

Table 6.5 Species distribution of some major seawater constituents.

Ion	Concentration ($mol l^{-1}$)	Free ion (%)	With sulphate (%)	With bicarbonate (%)	With carbonate (%)
Ca^{2+}	0.0104	91	8	1	0.2
Mg^{2+}	0.0540	87	11	1	0.3
Na^+	0.4752	99	1.2	0.001	—
K^+	0.0100	99	1	—	—

Ion	Concentration ($mol l^{-1}$)	Free ion (%)	With Ca (%)	With Mg (%)	With Na (%)	With K (%)
SO_4^{2-}	0.0284	54	3	21.5	21	0.5
HCO_3^-	0.00238	69	4	19	8	—
CO_3^{2-}	0.000269	9	7	67	17	—

QUESTION 6.14 (a) Which ion is missing from Table 6.5 and why?

(b) Does this help to explain why a much smaller proportion of cations than anions occur as free ions, according to Table 6.5?

(c) Which ion pair seems to be the most abundant, and in what context have you already encountered it?

6.3.2 THE CARBONATE SYSTEM, ALKALINITY AND THE CONTROL OF pH

The calcium carbonate used by many planktonic organisms to form their hard parts (Figure 6.3) redissolves when the organisms die and sink into deep water, releasing calcium and carbonate ions back into solution:

$$Ca^{2+} + CO_3^{2-} \rightleftharpoons CaCO_3\,(\text{solid}) \tag{6.6}$$

Accordingly, should Ca:*S* ratios be greater in deep or in surface waters?

If calcium is extracted from surface waters and then returned to solution in deep waters, Ca^{2+} concentrations should be higher in deep than in surface waters. Calcium is a bio-intermediate element, but it is so abundant in seawater that its involvement in biological processes results in only small increases of the Ca:*S* ratio with depth (*cf.* Section 3.1).

Ocean surface waters are nearly everywhere supersaturated with respect to calcium carbonate, which raises the question of why spontaneous inorganic precipitation of calcium carbonate occurs only infrequently. The reason lies in the inhibiting effect of Mg^{2+} ions in the solution: much of the carbonate is in the form of $MgCO_3$ ion pairs (Table 6.5). It requires the intervention of marine organisms to precipitate the calcium carbonate. Calcareous skeletal material is made of either **calcite** or **aragonite**, both of which have the same chemical formula, $CaCO_3$, but different crystalline structure. The aragonite structure is thermodynamically less stable than that of calcite, so aragonite dissolves more readily than calcite.

The depth at which significant dissolution of calcareous skeletal material begins (i.e. the depth where the water has become significantly undersaturated) is called the **lysocline**. The depth at which all the carbonate has dissolved is called the **carbonate compensation depth (CCD)** (not to be confused with the compensation depth defined for oxygen production/consumption—Section 6.1.3). The lysocline and CCD are shallower for aragonite than for calcite, and unless otherwise specified the terms usually refer to calcite, because skeletal material is much more commonly formed of calcite than aragonite. It is rare to find aragonite remains in sediments below about 1–2km, and sediments below 4km seldom contain significant amounts of calcite debris.

Variations in depth of the lysocline are controlled by the chemistry of the water column (carbonate equilibria and pH, see below). Variations in the CCD are controlled partly by chemistry and partly by the rate of supply of calcareous material sinking from the surface. Because it is not easy to determine exactly how much material has been dissolved, both lysocline and CCD are depth *zones* rather than precisely defined levels.

Other things being equal, would you expect the CCD to be deeper or shallower beneath areas of high biological production compared with areas of low production?

High biological production results in large populations of organisms and a high rate of supply of calcareous skeletal material to deep water when the organisms die. A heavy 'rain' of carbonate debris will reach greater depths before it all dissolves than would a meagre supply of calcareous material sinking from a region of low biological production. So the CCD tends to be depressed beneath areas of high biological production.

The behaviour of calcium carbonate is different from that of silica, which has no compensation depth. Indeed, the solubility of silica decreases as temperature falls; and so although silica does continue to dissolve as it sinks (Figure 6.1), much siliceous debris reaches the deep ocean floor (*cf.* Section 6.1.2).

We now need to look at the behaviour of carbon in more detail. The C:*S* ratio is a good deal more variable than the Ca:*S* ratio (Sections 3.1 and 4.3.4), but in general it changes in the same direction: it is lower in surface than in deep waters.

Why might that be?

The ratio is influenced both by the formation and subsequent dissolution of calcium carbonate skeletal material; and, quantitatively more important, by the formation of organic tissue during primary production in surface waters and its subsequent decomposition as it sinks from the surface. Figure 6.14 shows typical profiles for total dissolved carbon, expressed as ΣCO_2 (where Σ is capital sigma, and denotes 'sum of'). The profiles illustrate how carbon in its various forms (see below) is the least conservative of the major dissolved constituents with obvious bio-intermediate character.

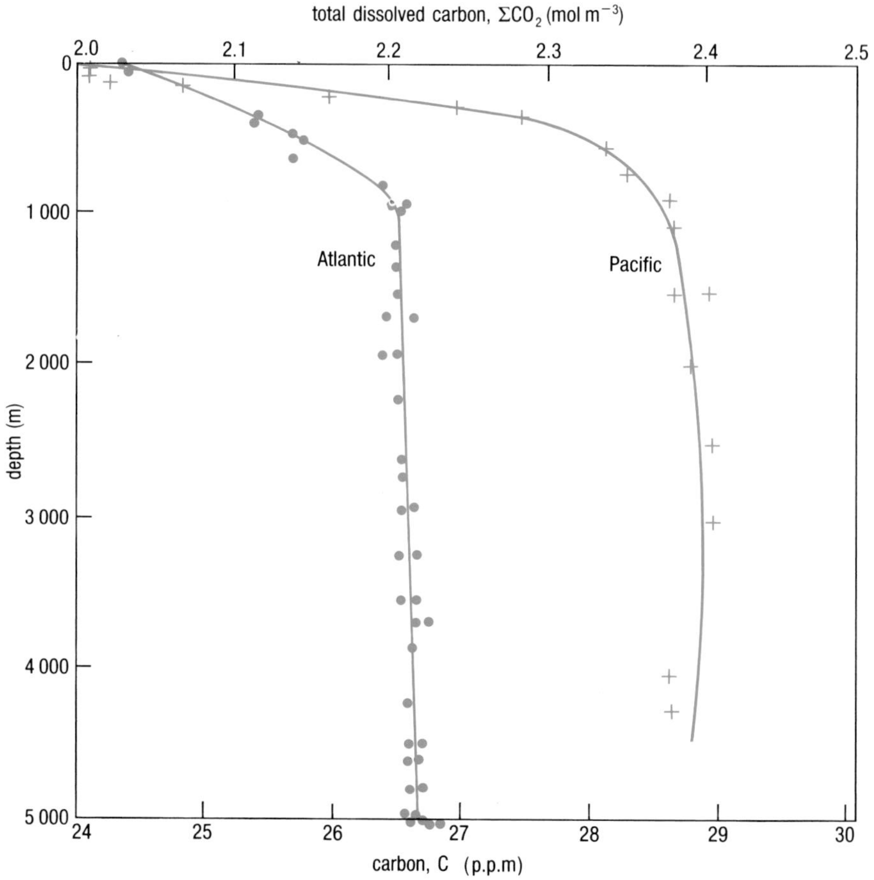

Figure 6.14 Variation with depth of total dissolved carbon expressed in mol m^{-3}, as well as p.p.m. of carbon, in the Atlantic at 36°N, 68°W (dots), and the Pacific at 28°N, 122°W (crosses). *Note:* to convert mol l^{-1} to mol m^{-3}, simply multiply concentration by 10^3, e.g. 2×10^{-3} mol l^{-1} becomes 2 mol m^{-3}.

Important: Because of reaction 6.2, CO_2 *as gas* is present in very small amounts in seawater: only $0.23\,ml\,l^{-1}$ at 24°C and atmospheric pressure. This is some 200 times less than the amount shown in Figure 6.4. Moreover, the concentration of CO_2 *as gas* increases only slightly as pressure increases with depth, as the concentration of total dissolved carbon, ΣCO_2, increases (Figure 6.14). So, when you read or hear about CO_2 concentrations in seawater, you should realize that this is simply a convenient shorthand way of describing the concentration of total dissolved carbon—and, in fact, CO_2 happens also to be the most convenient form in which to analyse total dissolved carbon. To be sure, the profile of increasing ΣCO_2 with depth in Figure 6.14 is largely the result of the production of CO_2 in respiration and decomposition of organic matter (reaction 6.3 goes to the left). But the CO_2 produced is not liberated as bubbles of gas, because as soon as it forms it combines with water in reaction 6.2. By far the most important components in that reaction are HCO_3^- and CO_3^{2-}, and it is to the relationships between these two ions that we now turn.

Alkalinity

Carbon occurs as several species in solution: CO_2 gas, H_2CO_3, HCO_3^-, CO_3^{2-}, as well as carbon combined in organic molecules. HCO_3^- and CO_3^{2-} are quantitatively by far the most important of these, and for simplicity, the others will be ignored here.

Important: The manipulation of chemical equations which now follows is much less frightening than it looks. The successive stages are carefully explained, step by step, so that you can follow them, even if your experience of chemistry is limited. The square brackets enclosing chemical formulae are the conventional symbol for concentration. So, for example $[CO_3^{2-}]$ simply means the concentration of carbonate ion. In this discussion we shall use concentrations expressed in $mol\,m^{-3}$.

It is difficult to measure the relative proportions of bicarbonate and carbonate ions directly, but their combined concentrations can easily be determined. Titration of seawater with acid neutralizes the negative charges, with the formation of water and CO_2, which leaves the solution as gas.

$$HCO_3^- + H^+ \rightarrow H_2O + CO_2\ (gas) \qquad (6.7)$$

$$CO_3^{2-} + 2H^+ \rightarrow H_2O + CO_2\ (gas) \qquad (6.8)$$

Why does it require two moles of hydrogen ion to neutralize the charge on one mole of carbonate ion?

Because the carbonate ion carries two negative charges. Only one mole of hydrogen ion is required to neutralize a mole of the singly charged bicarbonate ion.

The amount of acid used in the titration provides a value for the **alkalinity, *A*,** of seawater.

It is, in effect, the total amount of hydrogen ions required to neutralize the negative charges on the carbonate and bicarbonate ions, in equations 6.7 and 6.8. One mole of hydrogen ions is required to neutralize one mole of negative charge. So, the alkalinity is the combined molar concentration of carbonate and bicarbonate ions, expressed in 'charge-equivalent' terms.

For our purposes, alkalinity can be defined as the total charge carried by the carbonate and bicarbonate ions, expressed in terms of molar concentrations.

From the description of alkalinity just given, it can be expressed as:

$$A=[HCO_3^-]+2[CO_3^{2-}] \tag{6.9}$$

Why is the concentration term for carbonate ion doubled in equation 6.9?

It is doubled because (from equation 6.8) two hydrogen ions are required to neutralize the two negative charges on the carbonate ion.

If the gaseous CO_2 produced in the titration (equations 6.7 and 6.8) is collected and measured, its amount provides a measure of the total dissolved carbon in the seawater sample (remember that we are neglecting the small contributions of other forms of dissolved carbon):

$$[\Sigma CO_2]=[HCO_3^-]+[CO_3^{2-}] \tag{6.10}$$

Why is the concentration term for carbonate ion *not* doubled in equation 6.10?

It is not doubled because one mole of carbonate ion produces one mole of molecular CO_2, even though it requires two moles of hydrogen ion to do so (equation 6.8).

If we now re-arrange equation 6.9, we get:

$$A-[CO_3^{2-}]=[HCO_3^-]+[CO_3^{2-}] \tag{6.9a}$$

Compare the right-hand sides of equations 6.9a and 6.10. Do you see any resemblance?

The two right-hand sides are identical. It follows that the left-hand sides of these two equations must be equal, and so we can write:

$$A-[CO_3^{2-}]=[\Sigma CO_2] \tag{6.11}$$

A more convenient form of writing that equation is to have the two measured quantities on the same side:

$$A-[\Sigma CO_2]=[CO_3^{2-}] \tag{6.12}$$

The outcome of all this is to show that by measuring alkalinity and total dissolved carbon, equation 6.12 can be used to determine the carbonate ion concentration; and substitution of that value into equation 6.9 or 6.10 gives the concentration of bicarbonate ion. The reason for doing this is that the bicarbonate:carbonate ratio provides the main control on the pH of seawater.

Before we turn to that, how would you expect the alkalinity of seawater to change with depth?

We know that $[\Sigma CO_2]$ increases with depth (Figure 6.14), and alkalinity must therefore also increase with depth (equations 6.9 and 6.10).

The control of pH

pH is a measure of the concentration of hydrogen ions in a solution (see also the Appendix):

$$pH = -\log_{10}[H^+] \qquad (6.13)$$

In seawater, pH is mostly in the range 8.0 ± 0.2 and variations of pH are controlled chiefly by a component of reaction 6.2:

$$HCO_3^- \rightleftharpoons H^+ + CO_3^{2-} \qquad (6.14)$$

The reaction is very rapid and seawater can be assumed to have an equilibrium mixture of the three ions. When reaction 6.14 is at equilibrium, we can write:

$$K = \frac{[H^+][CO_3^{2-}]}{[HCO_3^-]} \qquad (6.15)$$

where K is the equilibrium constant.

If we re-arrange equation 6.15:

$$[H^+] = K\frac{[HCO_3^-]}{[CO_3^{2-}]} \qquad (6.16)$$

Equation 6.16 shows that the ratio of the concentration of HCO_3^- ions and CO_3^{2-} ions must control the hydrogen ion concentration and hence pH.

It is a matter of observation that A increases with depth by a greater amount than $[\Sigma CO_2]$. The effect of this on the way pH changes with depth is illustrated in Question 6.15.

QUESTION 6.15 The alkalinity of a sample of surface water is $2.35\,mol\,m^{-3}$ and its $[\Sigma CO_2]$ is $2.0\,mol\,m^{-3}$; the same quantities for a sample of deep water from the same location are $2.55\,mol\,m^{-3}$ and $2.4\,mol\,m^{-3}$. Use equations 6.12, 6.10 and 6.16 (in that order) to work out the pH for these two samples. Use a value of 1.0×10^{-9} for the equilibrium constant K.

If you need to refresh your memory on how to work out pH from $[H^+]$ and vice versa, consult the Appendix.

At this point, an obvious question presents itself: why not measure pH directly? The reason is that seawater is, chemically speaking, a concentrated solution (Section 6.3.1), and so reliable values of pH are not easily obtained by direct measurement.

The relationships outlined above have been simplified in order to establish some basic principles of seawater chemistry. As usual, reality is not quite so simple. For example:

1 We have ignored the other forms of dissolved carbon as being quantitatively insignificant, but they cannot be neglected when accurate measurement and calculation are required.

2 The alkalinity defined above should (strictly speaking) be called *carbonate alkalinity,* because other ions, notably boron species such as $H_2BO_3^-$, contribute to the total alkalinity of seawater, and it is in fact the total alkalinity that is measured by titration.

3 The equilibrium constants for the component reactions of the carbonate system (e.g. equation 6.15) are not strictly 'constant', but

change with temperature and pressure. None of these complications affects the basic principles, however, and the following generalization expands on what was said in Section 6.1.3, to the effect that: 'As CO_2 concentrations increase, so pH falls...':

> As a general rule, the greater the $[\Sigma CO_2]$, the smaller the value of $(A - [\Sigma CO_2])$, the greater the value of $[HCO_3^-]/[CO_3^{2-}]$, the higher the value of $[H^+]$, the lower the pH, and the more acid the water.

The corollary is that where $[\Sigma CO_2]$ is low, pH is high. For example, one of the few places where inorganic precipitation of calcium carbonate occurs is on the Bahamas Banks, where the sea is shallow and warm and the salinity is high (greater than 37). The warmer and more saline the water, the lower the solubility of gases, including carbon dioxide. In these conditions, the term $A - [\Sigma CO_2]$ in equation 6.12 will be large.

What will that mean for the value of $[CO_3^{2-}]$?

The concentration of carbonate ions will also be large, and it often rises sufficiently for the water to be so supersaturated with respect to $CaCO_3$ that the inhibiting effect of the $MgCO_3$ ion pair is overcome, and small crystals of calcium carbonate (in the form of aragonite) are precipitated. Equations 6.9 to 6.16 thus help to explain what might seem at first sight a paradox: where $[\Sigma CO_2]$ is high, calcium carbonate is more likely to dissolve; and where $[\Sigma CO_2]$ is low, calcium carbonate is more likely to precipitate.

The concepts introduced in this Section are not easy, but they are important because of the central role of carbon in seawater chemistry and because of their relevance to the global CO_2 problem, discussed in Chapter 7. Before moving on, however, it is important to correct a potential misconception. From the definition of alkalinity and the discussion in this Section, it should be clear that alkalinity is *not* a measure of how 'alkaline'—as opposed to 'acid'—seawater is. Once this is realized, it is much easier to come to terms with another apparent paradox which takes some getting used to, i.e. that both alkalinity and 'acidity' change in the same direction: where $[\Sigma CO_2]$ is high, so is alkalinity, and so is 'acidity' (low pH). Conversely, where $[\Sigma CO_2]$ is low, so is alkalinity, and so is 'acidity' (high pH).

6.3.3 MINOR AND TRACE ELEMENTS

The third column of Table 6.6 shows the predicted upper limits of concentration for some minor and trace elements in seawater. These are based on the measured solubilities of the least soluble compound likely to be formed by these elements in the seawater solution.

Compare the third and fourth columns of Table 6.6. Do the data suggest that seawater is supersaturated or undersaturated with respect to the elements listed in the Table?

Except for lanthanum (La), all the elements listed in Table 6.6 have smaller concentrations in the last column than the third column—especially mercury. Concentrations are significantly less than predicted concentrations. In other words, all the elements in Table 6.6, other than lanthanum, are undersaturated in seawater. Similar results are of course obtained for compounds formed by the major constituents—except

Table 6.6 Potential solubility controls on the concentrations of trace elements in seawater.

Element	Least soluble compound under seawater conditions	Saturation concentration under seawater conditions ($mol l^{-1}$)	Upper limit of observed concentration ($mol l^{-1}$)
Lanthanum	$LaPO_4$	8×10^{-12}	2×10^{-11}
Thorium	$Th_3(PO_4)_4$	2×10^{-12}	2×10^{-13}
Cobalt	$CoCO_3$	3×10^{-7}	6×10^{-9}
Nickel	$Ni(OH)_2$	6×10^{-4}	10^{-7}
Copper	$Cu(OH)_2$	2×10^{-6}	5×10^{-8}
Silver	$AgCl$	6×10^{-5}	3×10^{-9}
Zinc	$ZnCO_3$	2×10^{-4}	2×10^{-7}
Cadmium	$CdCO_3$	10^{-5}	10^{-9}
Mercury	$Hg(OH)_2$	80	8×10^{-10}
Lead	$PbCO_3$	3×10^{-6}	2×10^{-10}

$CaCO_3$ (Section 6.3.2)—so we do not expect NaCl, KCl and $MgSO_4$, for example, to be precipitated from seawater unless concentrations are greatly increased by evaporation. This is consistent with the long residence times and relatively high 'percentage in solution' values for the major constituents (Tables 6.3 and 6.4). But for the trace elements we are left without an explanation for their low 'percentage in solution' values and short residence times (Table 6.3 and Figure 6.11). They are undersaturated in seawater, and their short residence times mean that they are removed quickly from the seawater solution—they move rapidly from source to sink. But how? Some possible mechanisms of minor and trace element removal are considered below. They should not be considered as being mutally exclusive.

1 Minor and trace elements need not, in any case, form their own insoluble compounds such as those in Table 6.6. It is more likely that they will follow more abundant elements into precipitated phases, e.g. cobalt in the iron mineral goethite to give (Fe,Co)OOH, or lead in manganese oxide to give $(Mn,Pb)O_2$. The result is that trace elements can be removed from solution at concentrations less than the 'saturation concentrations' in Table 6.6.

2 **Adsorption** of metal ions (or ion pairs or complexes, *cf.* Figure 6.13) on to the surfaces of particles of both organic (detritus) and inorganic (clay minerals, hydroxides) origin, as they sink through the water column, is another mechanism for removing trace elements from the seawater solution—a process also called *scavenging*. Adsorption results from the mutual attraction between the charges on ions in solution and small residual surface charges on the particles.

Just how effective particle scavenging is as a mechanism for rapidly removing dissolved constituents from seawater was dramatically illustrated in the aftermath of the Chernobyl nuclear accident of April 1986. Sediment traps in the Mediterranean, North Sea and Black Sea recorded large increases in radionuclides at depths of 200m or more, within days of the arrival of the 'radioactive cloud' overhead. The research showed that the radionuclides were rapidly adsorbed on to particles which were in turn ingested by zooplankton and aggregated into faecal pellets. These sank at rates of several tens of metres per day (*cf.*

Question 6.2), taking the radionuclides with them—the radionuclides evidently passed rapidly through the digestive tracts of the animals. The radioactivity was removed from the surface to the sea-floor in a matter of weeks, or a few months at most. The radionuclides included isotopes of ruthenium (Ru), caesium (Cs) and cerium (Ce), the majority with half-lives of weeks, so they would probably not present a significant danger to the bottom-living animals (benthos). The exception is ^{137}Cs, with a half-life of 30 years, which could have longer term effects where concentrations were high.

4 **Oxidation–reduction equilibria:** A change in the valency state of an element upon oxidation or reduction can greatly affect its solubility. Many elements are able to participate in oxidation–reduction (or **redox**) reactions because they have variable valency. Important examples in marine chemistry are many of the transition elements (e.g. iron, manganese, chromium, vanadium), as well as nitrogen, sulphur and iodine. The degree of oxidation–reduction (or the redox state) of natural waters will control the solubility equilibria for several of these elements. For example, the trivalent form of iron (iron(III)) is much less soluble than the divalent form (iron(II)), so in water sufficiently oxidizing for trivalent iron to be dominant the amount in solution will be small because most of the iron will exist as iron(III) in solid phases such as $Fe(OH)_3$, or the mineral goethite, FeOOH. Under more reducing conditions, the dominant valency state for iron will be +2, and much higher concentrations of soluble iron (as Fe^{2+}) are to be expected. Two other examples are cobalt and manganese, which are present in seawater as Co^{2+} and Mn^{2+}, respectively, but can be oxidized to less soluble Co^{3+} and Mn^{4+} and precipitated as hydroxides or hydrated oxides.

Oxidation and reduction are formally defined as loss and gain of electrons, respectively. For example, in the reaction:

$$Fe^{3+} + e^{-} \rightleftharpoons Fe^{2+} \qquad (6.17)$$

the addition of an electron to iron(III) reduces it to the iron(II) state. Conversely, divalent iron(II) loses one electron and is oxidized to trivalent iron(III).

The availability in the water of species that can accept or donate electrons determines whether reaction 6.17 moves to the right or the left, that is, whether iron is present in its (less soluble) iron(III) or its (more soluble) iron(II) state.

An environment that is a net donor of electrons is therefore oxidizing; and one that is a net accepter of electrons is reducing.

QUESTION 6.16 (a) Is seawater normally an oxidizing or reducing medium?

(b) Is the natural form of iron in normal seawater the more soluble or less soluble form?

6.3.4 BIOLOGICAL CONTROLS ON CONCENTRATIONS OF DISSOLVED CONSTITUENTS

We have seen that biological activity has rather little effect on the concentrations of major constituents in seawater—even the important carbonate reactions described in Section 6.3.2 are associated with very small changes in total concentration. For minor and trace elements, however, it is a different story.

Shellfish have long been known to concentrate trace metals, with enrichment factors of many thousands (Table 6.7). It follows that where industrial effluents discharged at coastal sites contain metal concentrations above those of normal seawater, levels in shellfish may be correspondingly higher too, and this could make them toxic to organisms at higher levels in the food chain—including humans.

Table 6.7 Enrichment factors* for the trace element composition of shellfish compared with the marine environment.

Element	Enrichment factors Scallop	Oyster	Mussel
Ag	2300	18700	330
Cd	2260000	318700	100000
Cr	200000	60000	320000
Cu	3000	13700	3000
Fe	291500	68200	196000
Mn	55500	4000	13500
Mo	90	30	60
Ni	12000	4000	14000
Pb	5300	3300	4000
V	4500	1500	2500
Zn	28000	110300	9100

$$\text{*Enrichment factor} = \frac{\text{weight of element per unit weight of organism}}{\text{weight of element per unit weight of seawater}}$$

Note: These data are for dry weight of whole soft tissue. Live weight (or net weight) enrichment factors are about ten times less.

Plankton usually concentrate trace elements more strongly than do organisms further along the food chains, but the enrichment mechanisms are complex. The ability to concentrate a trace constituent from seawater may be shared by a whole class of organisms but very often uptake is limited to a particular family or even a single species. For example:

1 Copper is incorporated in haemocyanin, the blood pigment of molluscs and crustaceans (*cf.* iron in haemoglobin).

2 Vanadium is concentrated by certain ascidians or sea-squirts (e.g. *Ascidea ceratoides*) that have an organic vanadium complex as a blood pigment, whereas other species show low enrichment factors for vanadium, but do concentrate niobium, its vertical neighbour in the Periodic Table.

3 Some sponges (e.g. *Dysidea crawshayi*) accumulate titanium whereas others of the same genus (e.g. *Dysidea etheria*) do not.

4 Accumulation of polonium (a radioactive decay product of uranium-238) in prawns of the genus *Gennadas,* can result in their receiving α-radiation doses that are twice the lethal limit for humans.

The following enrichment mechanisms have been suggested to occur either in isolation or collectively:

1 Ingestion of particulate suspended matter such as clays or organic particles which have scavenged minor or trace elements from seawater (Section 6.3.3). This will be most significant for filter-feeding organisms.

2 Ingestion of elements already concentrated in food material: plankton concentrate the trace elements and then species higher in the food chain

eat the plankton. Such progressive concentration has so far only been shown to occur for mercury and manufactured organic compounds such as DDT and PCBs (polychlorobiphenyls).

3 Complexing of metals with organic molecules. At the mucus surfaces of digestive glands or gills of many organisms there are large molecules such as glycoproteins, which can form complexes with metal ions.

4 Incorporation of metal ions into physiologically important systems (e.g. copper into haemocyanin, cobalt into vitamin B_{12}).

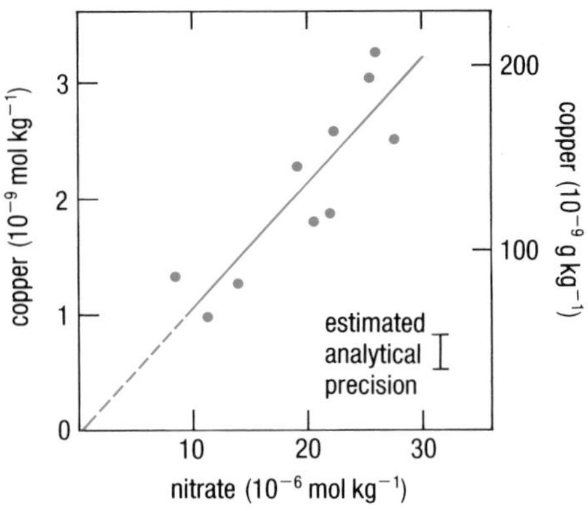

Figure 6.15 Plot of nitrate versus copper concentrations in water samples from the Antarctic Ocean, showing a clear covariance, with the extrapolated best-fit line through the data intersecting the origin.

The ability of seaweeds, especially *Laminaria,* to concentrate iodine, as well as sodium and potassium, has led in the past to the harvesting of the seaweed and extraction of these elements. (At present, seaweeds are collected on an industrial scale as a source of alginates which are used, amongst other things, as gelling and emulsifying agents in the food processing industry. This has nothing to do with their trace element content.)

Biological processes thus clearly influence the trace element composition of seawater, and a number of trace elements exhibit biolimiting or bio-intermediate behaviour. Two examples are shown in Figures 6.15 and 6.16.

QUESTION 6.17 (a) How does Figure 6.15 suggest that copper is a biolimiting micro-nutrient?

(b) How is it possible to infer from Figure 6.16 that nickel distribution in ocean waters may be biologically controlled?

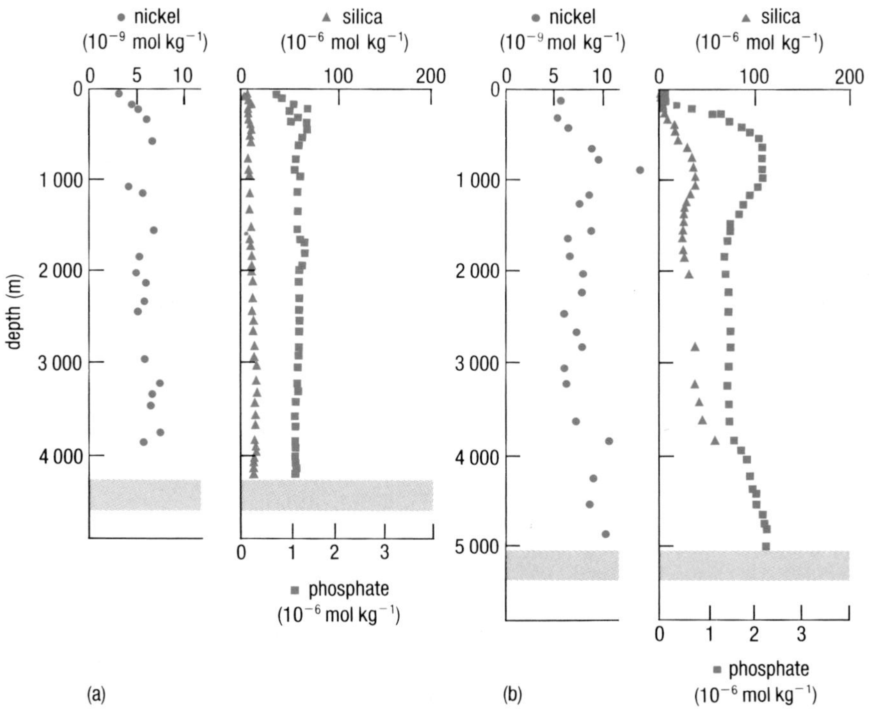

Figure 6.16 Profiles for nickel, phosphate and silica in (a) the North Atlantic and (b) the tropical Atlantic showing that nickel distribution generally follows that of phosphate and silica. Note that the nickel concentrations (dots) are in 10^{-9} mol kg^{-1}; phosphate (squares) and silica (triangles) are in 10^{-6} mol kg^{-1}.

The redox state of trace elements affects their behaviour in marine biological systems. For example, selenium is more readily available to marine organisms in its hexavalent form (SeO_4^{2-}) than in its tetravalent form (SeO_3^{2-}); trivalent arsenic (AsO_3^{3-}) is generally a good deal more toxic than the pentavalent form (AsO_4^{3-}).

Redox equilibria are of course not the only factor controlling speciation of elements in seawater and hence their availability to marine organisms. By way of another example, until the 1960s, mercury was thought to take part in marine biological processes only in inorganic forms, but in 1963 (the Minamata Bay tragedy in Japan) it was found that numerous cases of mercury poisoning, some fatal, were due to eating locally caught fish that had accumulated organic mercury compounds from industrial effluents. It is now known that mercury is more readily taken up by marine organisms when in the form of organic complexes (especially methyl mercury), than in its simple ionic forms (Hg^{+}, Hg^{2+}). The same is true of lead: marine organisms 'prefer' organic lead complexes (especially alkyl lead) to the simple ionic forms (Pb^{2+}, Pb^{4+}).

6.3.5 BIOLOGICAL ACTIVITY AS A SINK FOR TRACE ELEMENTS

Figure 6.17 summarizes the fate of the organic carbon fixed by photosynthetic primary production in the photic zone. About 90% of the organic matter forming the soft parts of phytoplankton is recycled above the thermocline, because of consumption by animals and bacterial decomposition of detritus and the products of excretion. Most of the remaining 10% is recycled as it sinks towards the sea-floor. Only a small fraction reaches the bottom, and most of that is consumed or decomposed by the deep-sea benthos (bottom-living organisms). Very little organic matter is preserved in the sediments.

The minor and trace elements taken up by marine organisms will in general share the same fate as the organic carbon, because they are

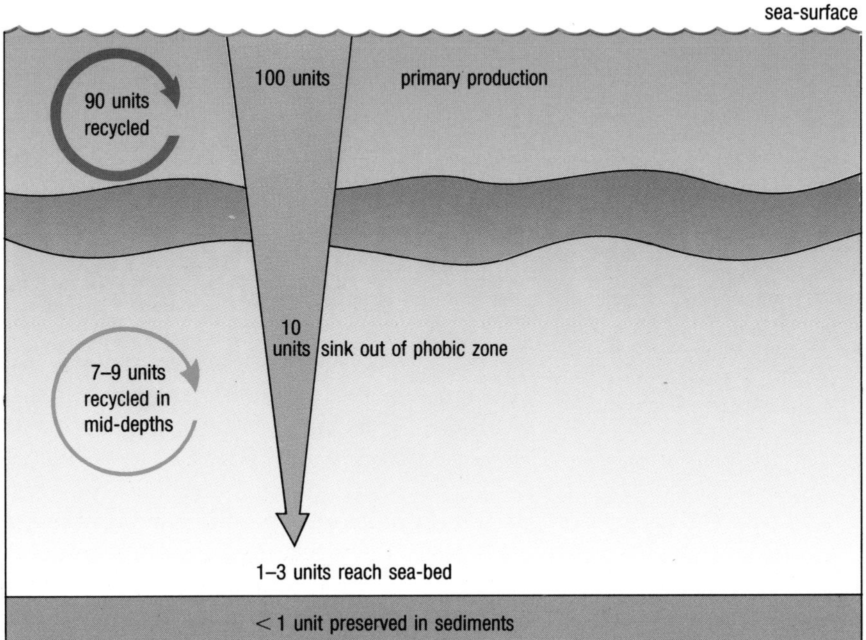

Figure 6.17 Sketch (not to scale) to illustrate the progressive decrease with depth of the organic carbon fixed by primary production in the photic zone.

concentrated mainly in the soft tissues. However, the concentrations of these elements in the sinking organic detritus probably increases progressively with depth. This can happen in two ways. First, the detritus becomes more 'refractory', as organisms feeding on it make use of the nutritious components and reject the rest. Secondly, the particles adsorb (scavenge) trace elements from solution as they sink through the water column (Section 6.3.3). Any minor and trace elements concentrated in skeletal material will partly be returned to deeper levels of the water column during re-solution at depth, and partly preserved in sediments.

The activities of organisms may thus provide a sink for minor and trace elements that is as important as the 'inorganic' sinks outlined in Section 6.3.3. It can be a great deal more important in coastal regions of upwelling where there is high productivity. In such conditions, the supply of dissolved oxygen is insufficient to oxidize fully the enhanced 'rain' of organic particles, and metal-containing organic compounds may not be decomposed. Their burial in the sediment becomes a permanent sink for the trace elements that they contain. Trace element enrichment in organic sediments (e.g. black shales) indicates that this process can be of considerable local importance.

Where does all this discussion of sinks for minor and trace elements leave the concept of the steady-state ocean?

It is important to bear in mind the analogy of the bottling factory and its conveyor belts. Constituents that are removed from the seawater solution and preserved in sediments (Figure 6.17) are, of course, replenished from the sources we have described earlier, namely river input, atmospheric fall-out and hydrothermal activity.

Although a principal objective of this Chapter has been to describe the behaviour of seawater itself, it must not be forgotten that there are also exchanges between seawater and the rocks and sediments on the sea-floor. Exchanges between atmosphere and ocean involve not only dissolved gasses but also salts and organic matter in aerosols, as well as heat and water (Section 2.2.1).

Exchanges at ocean boundaries have been going on ever since the Earth acquired its atmosphere and oceans, and in the concluding Chapter we look briefly at some aspects of the evolution of the Earth's fluid envelope.

6.4 SUMMARY OF CHAPTER 6

1 There are eleven major dissolved constituents of seawater, with concentrations greater than 1 part per million by weight (1×10^{-6}). The remainder are minor and trace constituents, the boundary between the two being about 1 part per billion by weight (1×10^{-9}). Most of the known chemical elements have been found in the seawater solution; it is likely that all are present and will eventually be detected. The boundary between what constitutes dissolved and particulate matter is chosen for practical reasons at 0.45 µm, but this does not always separate very fine colloidal particles from material truly in solution. Particulate matter (the seston) can remain in suspension for long periods because of turbulence.

2 Phosphate and nitrate are minor constituents and essential nutrients. They are extracted from surface water by photosynthesizing plankton to make organic tissue. They are totally depleted in surface waters where biological production is high, and are known as biolimiting constituents—they limit production because when they are exhausted, production ceases. When the organisms are consumed or when they die and decompose, the nutrients are returned to the water column. The molar ratio of N to P in both seawater and organic tissue is about 15:1. Silica is also a biolimiting nutrient, but is used only to make the hard parts of some planktonic organisms. The skeletal remains dissolve only slowly as they sink into deep water after death, and can accumulate in sediments on the sea-floor.

3 Carbon is essential to all life, but is so abundant in seawater that its involvement in biological production makes only a small difference to its concentration. Calcium is used to make calcium carbonate skeletons and shells, but like carbon it is so abundant that its concentration is little affected. Carbon and calcium are bio-intermediate constituents. Bio-unlimited constituents are those whose concentrations are unaffected by biological activity.

4 Carbon dioxide is the most abundant dissolved gas in seawater, followed by nitrogen, oxygen and argon, which dissolve in proportion to their atmospheric partial pressures. CO_2 forms carbonic acid with water, which dissociates to bicarbonate and carbonate ions, and these are the main form of dissolved carbon in seawater. Concentrations of gases may be given by weight or by volume ($\mathrm{ml\,l^{-1}}$), which are numerically not very different, because gas densities approximate to $1\,\mathrm{g\,l^{-1}}$. The solubility of gases decreases with increased temperature and salinity, and increases with pressure. Diffusion rates across the air–sea interface are increased in stormy weather, and dissolved gases are carried to deeper levels mainly by turbulent mixing.

5 Oxygen is supersaturated in surface waters. The compensation depth at the base of the photic zone can be defined as the depth at which the amount of oxygen used in respiration is equal to the amount of oxygen liberated by photosynthesis. Below the photic zone, respiration uses up available oxygen and an oxygen minimum layer develops at a depth of a few hundred metres. Deep water is richer in oxygen because of cold well-oxygenated water sinking in polar regions.

6 CO_2 concentrations increase with depth because CO_2 is used during photosynthesis and released again during respiration, and the solubility of CO_2 is increased by increased pressure. CO_2 concentrations are an important factor in controlling the pH of seawater, which is mostly within the range of 8.0 ± 0.2, being greater at the surface (less acid) than at depth (more acid). Many minor gases in seawater are produced by biological activity and are supersaturated in surface layers, so they have a net flux from sea to air.

7 Rainwater is a dilute version of seawater, because aerosols carry marine salts into the atmosphere where they provide nuclei for rain formation. The dissolved constituents in river water result from rock weathering and are dominated by calcium and bicarbonate ions, whereas seawater and rainwater are dominated by sodium and chloride ions. The chloride and much of the sodium in river water come from recycled sea salt. Sodium balance calculations show that most of the dissolved

constituents of seawater can be accounted for by rock weathering, but some cannot, notably chloride, bromide and sulphate. These are excess volatiles whose main source is probably volcanic gases. Hydrothermal activity in the ocean basins is an important additional source of some constituents of seawater.

8 The ocean is generally believed to be in a chemical steady state: rates of input and removal of dissolved constituents are in long-term balance. Residence times of dissolved constituents range from about 100 million years down to 1000 years or less, and there is a very rough correlation between concentration and residence time. The residence time of water in the oceans is about 4000 years and the stirring (or mixing or turnover) time is of the order of 500 years.

9 Dissolved constituents in seawater are mostly in ionic form, and all ions are surrounded by a hydration sphere. Most ions form ion pairs or ionic complexes, in which the hydration spheres are more or less merged. Several of the major constituent ions form ion pairs; two of the most abundant pairs are $MgSO_4$ and $MgCO_3$.

10 Some marine organisms use calcium carbonate to form their hard parts, which redissolve when the organisms die and sink into deep water. The depth at which dissolution begins is called the lysocline; the depth at which no calcium carbonate remains is called the carbonate compensation depth or CCD. $CaCO_3$ may be in the form of the less common and less stable aragonite, or the more abundant and more stable calcite. The lysocline and CCD are deeper for calcite than for aragonite. The concentration of calcium is greater in deep than in surface waters. So is that of total dissolved carbon (ΣCO_2), partly because of the dissolution of calcium carbonate, but chiefly because of the decomposition of organic tissue through consumption and respiration. To a good first approximation:

$$[\Sigma CO_2]=[HCO_3^-]+[CO_3^{2-}] \tag{6.10}$$

11 Alkalinity (A) is the combined negative charge due to bicarbonate and carbonate ions in seawater, expressed as molar concentrations. It is determined by titration with acid. As $[\Sigma CO_2]$ increases more with depth than does A, then from:

$$A-[\Sigma CO_2] = [CO_3^{2-}] \tag{6.12}$$

the concentration of carbonate ions decreases with depth; and from:

$$[H^+] = K\frac{[HCO_3^-]}{[CO_3^{2-}]} \tag{6.16}$$

deep water is generally more acid (lower pH) than surface water (higher pH); and skeletal remains formed of calcium carbonate dissolve as they sink into deep water.

12 Minor and trace elements in seawater generally have short residence times. Speciation is important, because several elements have more than one valency state and so can occur in more than one form. Different forms have different solubilities, and some may be precipitated by redox reactions or co-precipitated with insoluble complexes of other elements. Many are involved in biological processes and become greatly enriched in the tissues of marine organisms, so that organic-rich sediments may also be rich in trace elements. Some are biolimiting micro-nutrients; others show bio-intermediate behaviour.

Now try the following questions to consolidate your understanding of this Chapter.

QUESTION 6.18 Classify each of the constituents in Figure 6.18 in the (a) biolimiting; (b) bio-intermediate; or (c) bio-unlimited category.

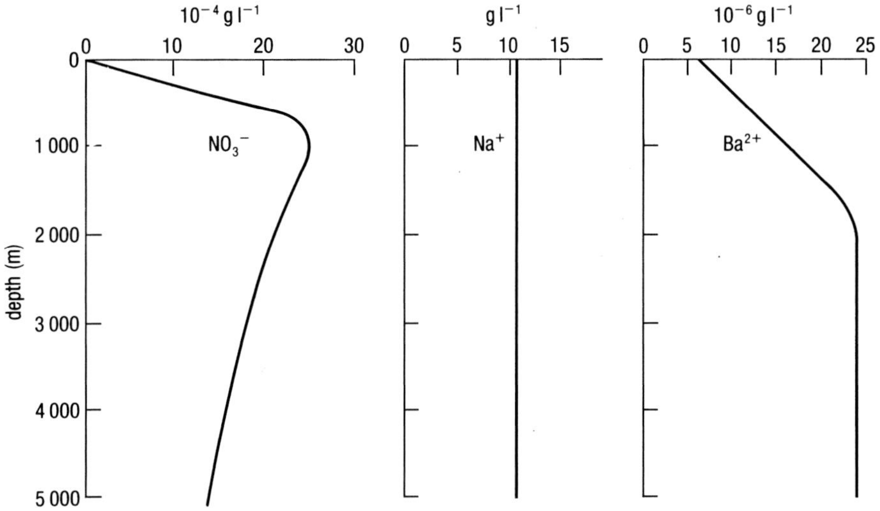

Figure 6.18 Concentration–depth profiles for three seawater constituents (for use with Question 6.18). Note that concentrations are normalized to a salinity of 35.

QUESTION 6.19 It is estimated that about 10^5 tonnes of dissolved manganese are delivered annually to the sea (either in dissolved ionic form or as colloidal particles) by rivers; but that some $4–6 \times 10^6$ tonnes of manganese are removed annually to deep-sea sediments (including manganese nodules). About 5×10^{11} tonnes of heated seawater circulate annually through the ocean crust at active spreading ridge axes and the average manganese content of this water is 10 p.p.m.

(a) How does the concentration of manganese in hydrothermal solutions compare with that in seawater?

(b) To what extent can the manganese in hydrothermal solutions account for the shortfall between river source and oceanic sink in the oceanic manganese budget?

(c) With the help of information in Table 6.1, calculate the residence time of manganese in the oceans. How does it compare with the values in Table 6.4 and Figure 6.11? Which is nearer to your answer, and why?

QUESTION 6.20 Under anoxic conditions, would you expect manganese to be in the form of soluble Mn^{2+} ions or insoluble oxide MnO_2?

QUESTION 6.21 Which of the following statements (a)–(g) are true, and which are false?

(a) Most of the nitrogen dissolved in seawater is in the form of nitrate ions.

(b) Dissolved gases are conservative constituents of seawater.

(c) The oxygen minimum layer is below the (oxygen) compensation depth.

(d) The shorter the residence time of a dissolved constituent in seawater, the greater its average concentration in river water relative to that in seawater (assuming rivers to be the main source).

(e) If Na^+ and Ca^{2+} ions have similar ionic size, then the Na^+ ion will have a larger hydration sphere relative to its size.

(f) The commercial extraction of salts from seawater by evaporation provides evidence that most major constituents are undersaturated in the seawater solution.

(g) A large influx of highly acid effluent (or a heavy downpour of acid rain) would tend to expel CO_2 from seawater.

CHAPTER 7 SEAWATER AND THE GLOBAL CYCLE

The role of the oceans in the global cycling of elements is illustrated in Figure 7.1 and may be summarized as follows:

1 Weathering of rocks supplies dissolved constituents to the oceans, aided both by the dissolution of mineral components in rocks of the oceanic crust during hydrothermal circulation, and by the supply of chloride, sulphate and other excess volatiles from volcanic gases.

2 Most dissolved constituents have residence times much longer than oceanic stirring times (Figure 6.11) and may be cycled repeatedly within the main body of the oceans, especially by participating in biological reactions. There can also be exchanges across the sea-bed and the air–sea interface.

3 Dissolved constituents are eventually removed from the seawater solution into sediments and rocks by the processes of reverse weathering. These include adsorption and co-precipitation of minor and trace elements, formation of skeletal material, preservation in anoxic environments, and reactions during hydrothermal activity.

4 Sediments and rocks are removed from the oceanic environment either by direct uplift above sea-level or by subduction into the Earth's mantle, at converging (destructive) plate margins. Uplift brings sediments and rocks directly back into the weathering environment; subduction eventually returns them to the crust via magmatic processes which form

Figure 7.1 Over periods of millions of years, the oceans behave as a well-mixed tank. The input and output processes are either external (Sun-driven: river flow, photosynthesis); or internal (Earth-driven: reactions at mid-oceanic ridges, uplift and subduction of oceanic crust).

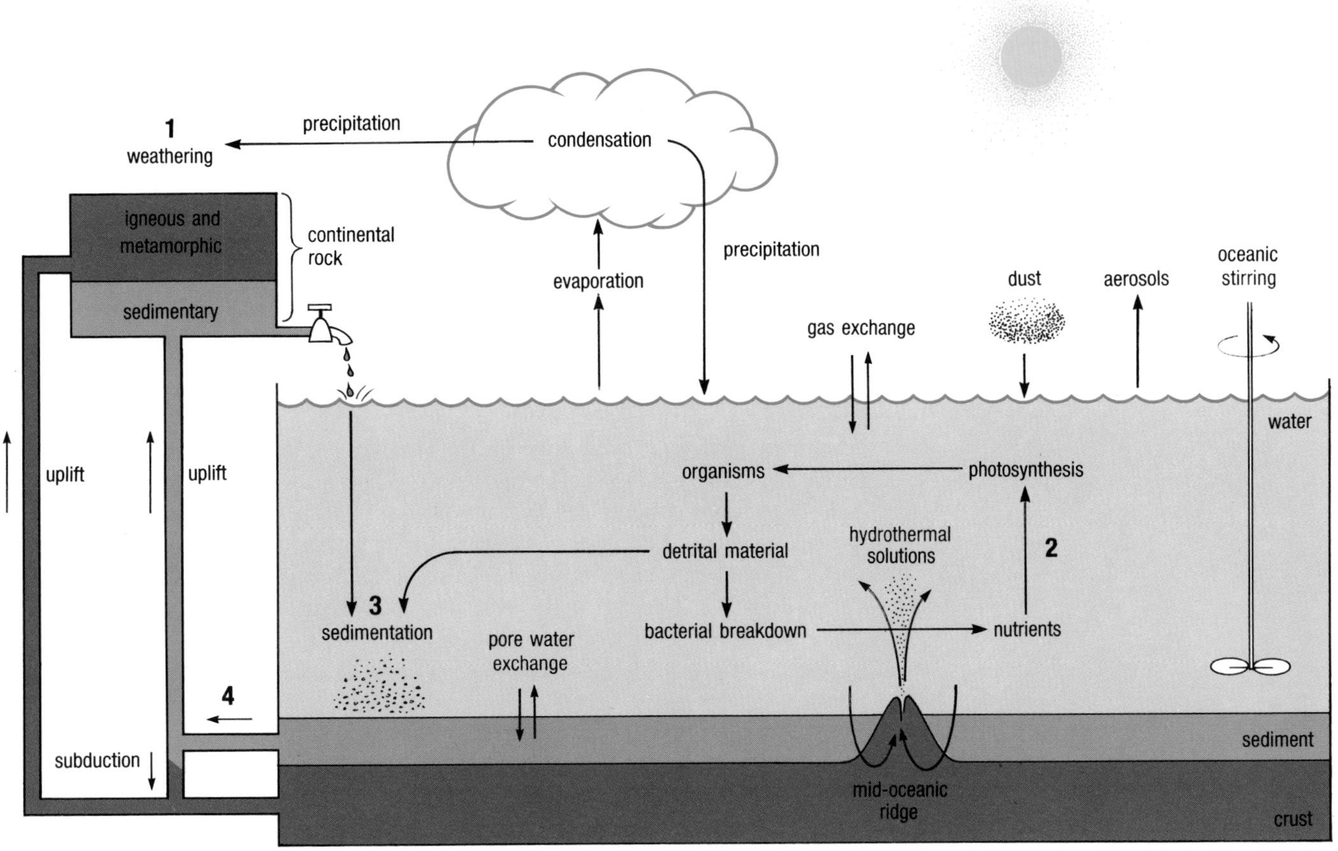

volcanic and intrusive rocks and release volcanic gases (the excess volatiles).

The excess volatiles are thus part of the global cycle. A small proportion is probably truly juvenile or primordial (i.e. derived from deep within the Earth, where it has resided from the beginning); but most of these volatiles have been circulating through the system summarized in Figure 7.1 for perhaps thousands of millions of years.

QUESTION 7.1 Carbon dioxide and methane are also volcanic gases. While a small proportion of both gases may be juvenile in origin, most is recycled. Can you suggest how this could happen?

If the composition of seawater has not changed significantly for at least several hundred millions years—and some authorities would extend this to more than a thousand million years—the chemical steady state is a fundamental characteristic of the oceans (Section 6.2.5). Rates of input and removal of most dissolved constituents are generally in balance. If it were not so, we would not be able to calculate residence times.

The broad correlations of Figure 6.11, along with other more subtle and complex relationships, suggest that the chemistry of the oceans has been controlled for most of geological time by some basic bio-geochemical rules, which have maintained the long-term stability of seawater composition. Geological evidence supports such a conclusion. Marine sedimentary and igneous rocks of all geological ages are compositionally very similar to their modern equivalents; this applies not only to the major components of limestones, sandstones and shales, but also to minor and trace elements. For example, evaporites older than 2500 million years retain evidence that sodium chloride was the principal salt to be precipitated. Proportions of elements such as copper, zinc and uranium in organic-rich (black) marine shales from ancient geological periods are similar to those in comparable sediments being deposited now in the Black Sea.

Long-term stability of composition does not mean eternal constancy, however, and there must have been some changes in seawater composition with time.

7.1 A SHORT HISTORY OF SEAWATER

The most significant influence on the surface environment of the early Earth must have been the composition of the atmosphere, which was dominated by nitrogen and carbon dioxide. There was some oxygen, because fossil soil profiles with ferric oxide horizons have been found in rocks from the Archaean era (more than 2500 million years old). Conditions were not strongly oxidizing, however, and it is possible that there was local reduction of nitrogen to ammonia. The primitive atmosphere probably also contained a certain amount of methane, another de-gassing product of the early Earth. Laboratory experiments in the 1950s showed that amino acids (the building blocks of protein) could have been naturally synthesized from these gases in solution in seawater. The necessary energy was probably supplied by lightning discharges and by ultraviolet radiation, which could penetrate the atmosphere to the

Earth's surface in the virtual absence of atmospheric oxygen and hence of an ozone layer. With the discovery of hydrothermal vents in the oceans during the 1970s, it was suggested that organic molecules necessary for the development of early life forms could also have originated in the deep sea. Hydrothermal vents would provide an ideal environment: plenty of hot water and an abundance of raw materials. The Earth was hotter early in its evolution than it is now, and the hydrothermal environment was more widespread. Whether life on Earth originated at the ocean surface or the deep sea-bed, the earliest life forms preserved in the fossil record are about 3.5 billion years old, and they were primitive blue–green algae, requiring sunlight for photosynthesis.

The composition of seawater on the early Earth must also have been different from what it is now. Dissolved constituents such as carbonate/bicarbonate and sulphate among the anions, and iron and manganese among the cations, must have been present in very different proportions. The concentration of HCO_3^- was probably second only to that of Cl^-, because sulphur would mostly have been in the form of relatively insoluble sulphide rather than soluble sulphate. The reduced forms of iron and manganese (Fe^{2+} and Mn^{2+}) are both readily soluble, and as they are major components of crustal rocks (Table 3.2) they would have been more abundant in seawater than they are now.

The ratio of atmospheric CO_2 to O_2 gradually decreased with time, as photosynthesizing organisms fixed carbon in organic tissue and released oxygen. A contributory factor was the precipitation of carbonate sediments (though organisms with calcareous skeletons did not evolve until about 600 million years ago).

There is an interesting speculation that life was unable to begin colonizing land surfaces until about 400 million years ago, because only then was there sufficient atmospheric oxygen for the ozone layer to be thick enough to cut off most of the harmful ultraviolet radiation from the Sun. Variations in the level of atmospheric oxygen during the last several hundred million years are not ruled out, however. Different authorities place rather wide limits on this variation: from as much as five times greater to as little as one-tenth of present levels since about 600 million years ago. With an oxidizing atmosphere and a progressive decrease in CO_2, both atmosphere and ocean approached their present compositions ever more closely. The temperature-buffering effect of the oceans (Section 2.1) has meant that the Earth's surface has nearly always been tolerable for life, but average surface temperatures have fluctuated appreciably. There is good geological evidence that the Earth's surface environment is normally characterized by ice-free poles and relatively gentle poleward temperature gradients. In short, the present-day Earth is in an atypical condition: we are probably still in the Pleistocene Ice Age, albeit enjoying the relative warmth of an interglacial interval. The previous major ice age occurred about 300 million years ago (the Permo-Carboniferous).

QUESTION 7.2 If higher average temperatures than now characterized the Earth's surface, and the poles were ice-free, what implications might such conditions have for the deep circulation?

It is worth amplifying the answer to Question 7.2: in an ice-free world, reduced oxygen levels would also mean less decomposition and more preservation of organic tissue in sediments. There would be less turnover of nutrients than now, because more organic tissue, containing the nutrients, was being preserved in sediments.

7.1.1 THE SPECIAL CASE OF CO_2

Carbon is the element which forms the basis of all life (Section 6.1.2) and carbon dioxide is the form in which it is used for photosynthetic primary production. Early in the Earth's geological history, CO_2 was about a thousand times more abundant than it is in the present-day atmosphere.

QUESTION 7.3 (a) Look back to Figure 6.4 and use it to estimate what the partial pressure of CO_2 was at an early stage in the Earth's history.

(b) What would that imply about its concentration in seawater, and why might the $[HCO_3^-]/[CO_3^{2-}]$ ratio in the early ocean have been greater than it is now? (A qualitative answer only is required.)

Some authorities estimate that early in the Earth's history, the solar luminosity was about 25 per cent less than it is now, so that the Earth received only about three-quarters of the present-day solar radiation. If that is correct, then solar luminosity must have increased with time, as atmospheric CO_2 has progressively decreased. As outlined in Section 2.1, carbon dioxide is the main contributor to the atmospheric greenhouse effect, along with water vapour. The more CO_2 in the atmosphere, the greater the effect.

If CO_2 levels had not fallen as solar luminosity increased, what would the Earth's surface be like now?

Very hot: rather like Venus is today. Unlike Venus, however, the early Earth was just far enough from the Sun for liquid water to exist at the surface, in rivers, lakes and seas. Carbon dioxide dissolved in the water, and the abstraction of carbon into organic tissue and sediments began soon afterwards. Much carbon is now locked up in crustal rocks, as well as in the biosphere and in fossil fuels (Table 7.1). Carbon continues to circulate through the global cycle, but the amount stored in the various reservoirs changes rather little. (Although the fossil fuel 'bank' is being rapidly depleted by human activities, this is a relatively small reservoir of carbon.)

Table 7.1 Amounts of carbon in various reservoirs (10^{12} tonnes CO_2 equivalent).

Reservoir	Approximate quantity
Atmosphere	2
Biomass (living matter)	30
Soils (organic carbon)	25000
Oceans and freshwater (in solution)	140
Carbonate sediments	150000
Fossil fuels (plus organic carbon disseminated in sediments)	27

The decline in atmospheric CO_2 has been progressive, but as in the case of atmospheric oxygen it may not have been regular. Figure 7.2 provides evidence that considerable short-term fluctuations of atmospheric CO_2 concentration have occurred during the past couple of hundred thousand years at least.

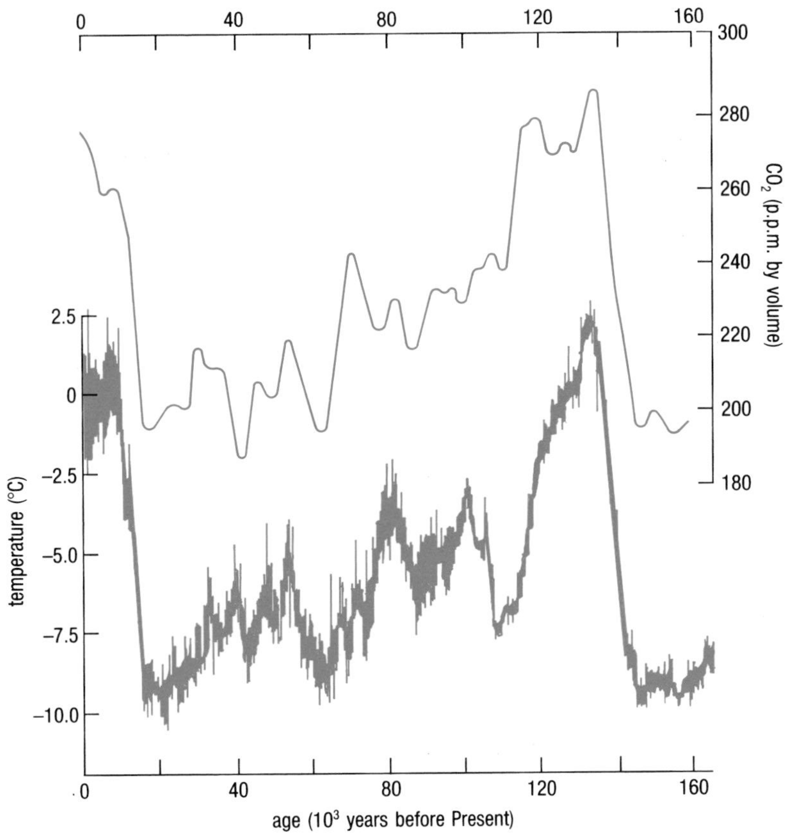

Figure 7.2 Variation with time in atmospheric CO_2 concentrations, determined from bubbles trapped in an ice core at Vostok in eastern Antarctica (blue curve); along with the atmospheric temperature at the surface, inferred from measurements of the deuterium/hydrogen isotopic ratio in H_2O (red curve).

Perhaps the most interesting feature of Figure 7.2 is not the fluctuations of atmospheric CO_2 themselves, but the strong correlation with surface temperature.

Is it a positive or a negative correlation?

The correlation is positive, that is, as temperature increases, so does the atmospheric CO_2 concentration. The deglaciation events commencing about 140000 years and 15000 years ago are particularly obvious. The fluctuations in Figure 7.2 are consistent with what might be expected from simple considerations of solubility and temperature alone. CO_2 is more soluble in cold than in warm water, and its atmospheric concentration should therefore be less during glacial (lower mean temperature) than interglacial (higher mean temperature) periods.

The full picture is more complex than that, however, and there has been considerable debate about whether changes in the atmospheric CO_2 concentration are a response to the temperature fluctuations or a cause of them. Several hypotheses have been proposed, relating changes in atmospheric CO_2 concentration to changes in biological productivity, in sea-level, and in the circulation of surface and deep current systems (including the relative importance of deep water mass formation in polar latitudes, and of upwelling regions, where biological production is high). As these factors are all interrelated, the resulting models are complex and no clear cause-and-effect relationships have emerged so far. It is reasonably certain, however, that varying CO_2 concentrations in the atmosphere are not the main cause of temperature fluctuations—they may reinforce climatic trends, once established, but are not likely to initiate them.

This is a suitable point at which to digress briefly into some external factors which influence the Earth's climate.

7.1.2 CLIMATE AND THE EARTH'S ORBIT

Seasonal variations in insolation were briefly described in Section 2.2. Figure 7.3 provides a more detailed picture of the way insolation at the Earth's surface varies with latitude and season.

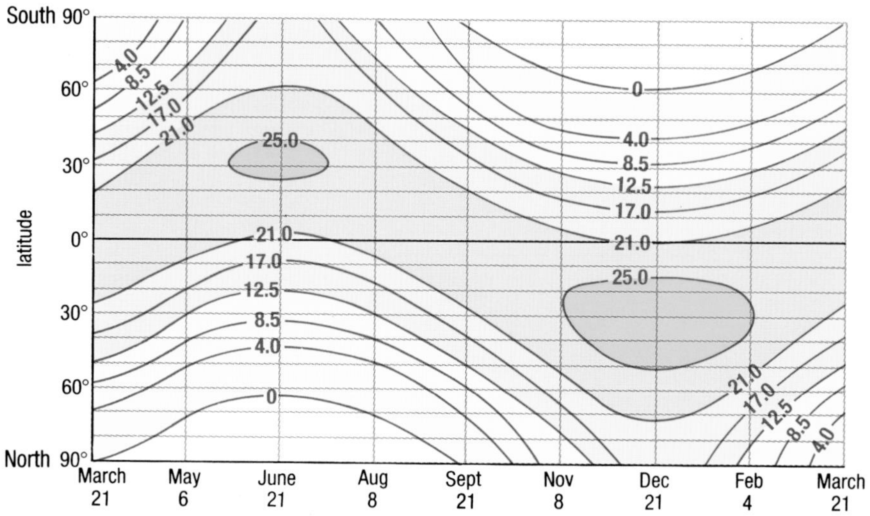

Figure 7.3 Seasonal variation of daily insolation (in 10^6 J m^{-2}) assuming 30% reflection from the top of the atmosphere (*cf.* Section 2.1). Values are highest in mid-latitudes because of long daylengths in summer.

QUESTION 7.4 Which parts of the Earth's surface receive the greatest insolation (a) at any one time, (b) on average over the year as a whole?

In answering Question 7.4, you may also have noticed that Figure 7.3 shows mid-latitudes in the Southern Hemisphere to receive more solar radiation than corresponding latitudes in the Northern Hemisphere, during their respective summer seasons. This is because the Earth is in an elliptical orbit around the Sun, so its distance from the Sun changes through the year. The positions on the orbit closest to and furthest from the Sun are called the **perihelion** and **aphelion** respectively.

So, in what part of the year is the Earth at perihelion, according to Figure 7.3?

As insolation is most intense during the southern summer, the Earth must be at perihelion during that season, i.e. close to the December solstice (*cf.* Figure 2.2)—early January in fact.

The geometry of the Earth's orbit round the Sun changes with time, and this affects the seasonal and latitudinal distribution of the insolation received by the Earth each year. We can identify three main changes in the orbital configuration:

1 Eccentricity: As a result of gravitational attraction from other planets in the Solar System (notably Jupiter), the Earth's orbit changes in shape from elliptical to nearly circular and back again, with a periodicity of about 110000 years.

In what major respect would Figure 7.3 look different if the Earth's orbit were nearly circular?

Insolation in both hemispheres would be about the same, because perihelion and aphelion become irrelevant when the orbit is circular. The degree of eccentricity of the Earth's orbit at present can be judged from the fact that the Sun–Earth distance is 1.521×10^8km at aphelion, 1.471×10^8km at perihelion.

2 Tilt: Figure 2.2 shows that the Earth's rotational axis is tilted with respect to the plane of its orbit round the Sun. The angle of tilt changes from about 22° to about 25° and back again, with a periodicity of about 40000 years, again because of the gravitational pull of the other planets. At present, the angle is about 23.5°.

3 Precession of the equinoxes: The seasonal position of the Earth in its orbit changes with a periodicity of about 22000 years.

We have just seen that the December solstice occurs when the Earth is close to its perihelion position. Looking back to Figure 2.2, about how long ago did the June solstice occur close to perihelion, and about how long will it be before it does so again?

The June and December solstices lie on opposite sides of the orbit, i.e. they are 180° apart. If the full cycle (360°) takes 22000 years, then the June solstice must have coincided with perihelion about 11000 years ago, and will do so again about 11000 years from now.

These three periodicities, *c.* 110000 years, *c.* 40000 years and *c.* 22000 years, are called **Milankovitch cycles**, after the scientist who first recognized that because they affect solar insolation patterns they should also be an agent of global climatic change. And so it has proved. Fluctuations with these periodicities can be discerned in temperature records determined from the oxygen isotope ratios measured in the calcareous skeletons of plankton preserved in deep-sea sediments. On Figure 7.2, the most prominent peaks are those corresponding to the *c.* 110000-year eccentricity cycle. There is evidence from deep-sea sediments that this approximates to the interval between major glaciation/deglaciation events in the Pleistocene Period of the last couple of million years.

However, the effects of the *c.* 110000-year orbital eccentricity cycle are generally considered insufficient by themselves to bring about the major changes of surface temperature and ice cover implied in compilations such as Figure 7.2. Interactions between the oceans, the ice-sheets and the atmosphere, and the characteristics of the ice-sheets themselves, as well as the effects of the Milankovitch cycles on surface insolation, must all be considered. A further complication is that the *c.* 110000-year eccentricity cycle, the effects of which seem so obvious on Figure 7.2, appears to be a characteristic of only about the last million years of the Earth's history. It seems to be a less obvious feature in records of earlier climatic fluctuations preserved in sediments, which are dominated by the *c*. 40000-year tilt cycle back to about two-and-a-half million years, the start of the Pleistocene ice age.

A fascinating suggestion to account for this difference is that the rapid tectonic rise of mountains in the Himalayas and western North America affected the circulation in the upper atmosphere in such a way as to strengthen the influence of the 110000 year eccentricity cycle—and it provides an example of how the solution of modern global problems requires information from many different fields of science.

This discussion has strayed some way from the CO_2 issue, not because it is irrelevant, but because the precise role of CO_2 as an agent of past climatic change is not yet known. It is an increasingly important field of research, however, because the progressive increase in atmospheric concentration of this gas has major implications for our future.

7.2 A LOOK AHEAD

Figure 7.4 shows how the content of CO_2 in the atmosphere has increased since the Industrial Revolution. In recent years, this increase has accelerated, due partly to greater industrial activity, and partly to greatly increased deforestation and land clearance for urban, industrial and agricultural development. Many people are now aware of the predicted greenhouse effect of this increase (Section 2.1): atmospheric and surface warming on a time-scale of decades, with consequences such as melting ice-caps and rising sea-levels. The trend seems to be under way already. Mean global surface temperature has increased by about 0.5°C since the late nineteenth century, and mean sea-level has risen some 10–15cm in the same period, partly because of melting ice, but also because of thermal expansion of the top few hundred metres of the seawater column. By about the year 2030, mean temperature and sea-level may have risen further by similar amounts or even more. Carbon dioxide is not the only 'greenhouse gas' (Section 2.1), but its effects are greatest because its concentration in the atmosphere is increasing at the greatest rate.

QUESTION 7.5 How do present-day atmospheric concentrations of CO_2 (Figure 7.4) compare with those of 130000 years ago (Figure 7.2)?

Figure 7.4 shows, however, that not all of the CO_2 released by human activities has stayed in the atmosphere. Some of the excess is probably used up by increased rates of terrestrial photosynthetic production; the remainder has dissolved in the oceans, where it is available for both greater primary production and increased carbonate precipitation.

The biosphere thus appears to be counteracting the artificial increase in atmospheric CO_2 by acting as a sink for some of it, and so buffering the greenhouse effect. Earlier you read that the principal cause of the progressive fall in the atmospheric CO_2:O_2 ratio was biological activity, removing CO_2 and releasing oxygen during photosynthesis. Relationships of this kind have led to the novel concept, which is beginning to gain wider acceptance, that the surface of our planet is actively maintained as a life-supporting environment by biological activity which acts as a feedback mechanism. This is the cornerstone of the **Gaia Hypothesis**, put

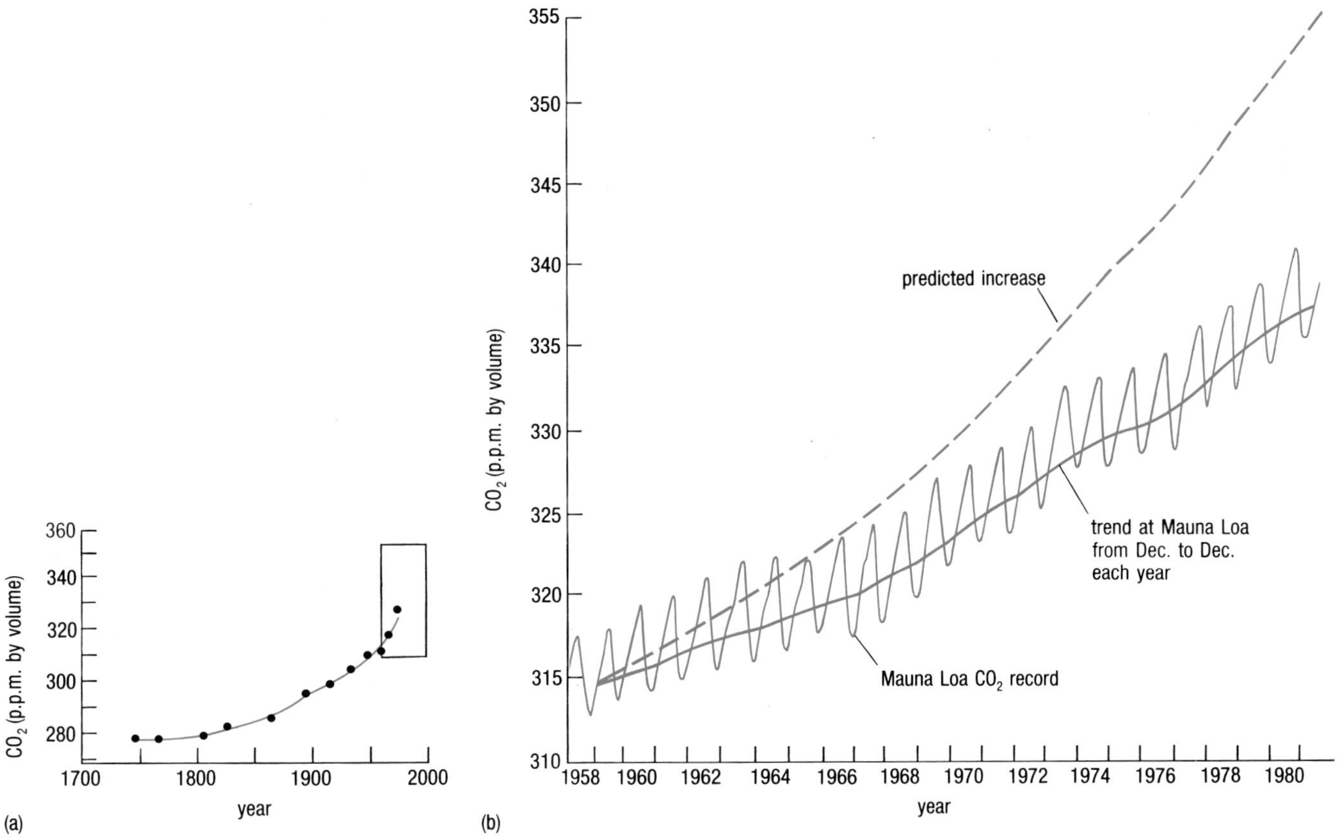

Figure 7.4 (a) Increase in atmospheric CO_2 since the start of the Industrial Revolution, determined from air trapped in Antarctic ice. (Rectangle at upper right is the post-1950 record, see (b).)

(b) Increase in atmospheric CO_2 measured at the Mauna Loa observatory, Hawaii, along with the increase predicted on the basis of fossil fuel combustion. Fluctuations are attributed to seasonal variations of primary biological production in the Northern Hemisphere.

forward in the early 1970s, but only coming to prominence in the mid-1980s. We shall end this Chapter with a brief and simple example of the way in which marine plankton may act to alleviate the greenhouse effect of rising CO_2 levels by a process other than simply 'soaking up' CO_2.

Large amounts of dimethyl sulphide or DMS ($(CH_3)_2S$) are produced by planktonic organisms, and there is a considerable net flux of this gas from sea to air (Table 6.2). In the atmosphere, it is rapidly oxidized to form sulphuric acid aerosols, which—together with dust and sea salt aerosols, Section 2.2.1—provide nuclei for the condensation of atmospheric water vapour into clouds and rain. Increases in atmospheric CO_2 concentrations lead to similar increases in oceanic surface waters and to increased marine productivity. This in turn leads to increased production of DMS, and hence to more cloud formation, aided by increased evaporation from the ocean surface into an atmosphere already being warmed up by the greenhouse effect.

What effect do clouds have on incoming solar radiation?

They reflect it back to space (Section 2.1). Although there is evidence that cirrus clouds in particular allow ultraviolet radiation to pass through but trap infrared, it seems likely that in general the more clouds there are, the less solar radiation reaches the sea-surface to warm it, the less infrared and microwave radiation it emits, and the less the greenhouse trapping of that radiation. The effect is helped by the fact that cloud cover is more extensive over oceanic than over continental areas.

This is a very brief and qualitative account of a process that has been researched and quantitatively modelled in some detail, and it is only one example among several that have been cited. It provides an additional mechanism for the maintenance of steady state conditions in the oceans over long periods of time; and we have seen in Chapter 6 that biological controls on seawater composition are just as important as inorganic chemical processes. It seems likely that the chemical composition of the seawater solution would have fluctuated much more widely if Earth had been a planet devoid of life.

7.3 SUMMARY OF CHAPTER 7

1 The oceans are maintained in a chemical steady state by the global cycling of elements through (i) weathering and solution; (ii) precipitation, re-solution and sedimentation; (iii) subduction, uplift, volcanic activity; and (iv) back to weathering. Sediments and rocks of the marine environment have changed little in composition through geological time, which suggests that seawater composition has not changed greatly and that elements have been cycled through it in similar proportions and amounts since early in the Earth's history.

2 Early seawater compositions may have been governed partly by high CO_2:O_2 ratios in the early atmosphere, with greater carbonate and lower sulphate concentrations and greater concentrations of reduced cationic species (e.g. Fe^{2+} and Mn^{2+}). CO_2:O_2 ratios decreased as carbon was fixed and oxygen released by photosynthesis.

3 The typical surface environment of the Earth is one of ice-free poles, higher average temperatures, more sluggish deep circulation and lower oxygen concentrations in seawater, with more organic matter preserved.

4 The high CO_2 concentrations of the early Earth's atmosphere probably resulted in more acid seawater, as the content of CO_2 was approximately at the level at which oxygen is today. The fall in CO_2 with time has been accompanied by increased solar luminosity, so that the greenhouse effect of atmospheric CO_2 has decreased with time, thus helping to maintain an equable surface temperature. The decline in atmospheric CO_2 has been progressive, but probably irregular.

5 Fluctuations in the concentration of atmospheric CO_2 are strongly correlated with surface temperature, at least over the last 150000 years, and probably much longer. Cycles of eccentricity of the Earth's orbit, changes in the angle of tilt of its axis of rotation and precession of the equinoxes (*c.* 110000, *c.* 40000 and *c.* 22000 years respectively—the Milankovitch cycles), are probably the major causes of climatic variation. Atmospheric CO_2 appears to reinforce trends of changing global temperature, but is probably not a major factor initiating them.

6 As CO_2 levels now increase again because of human activities, the greenhouse effect is predicted to lead to melting ice-caps and rising sea-levels. The effect is, however, being modified by the biosphere. Some CO_2 is used up by increased photosynthetic activity, and some is dissolved in the oceans.

7 The notion that biological activity helps to maintain the Earth's surface in a life-supporting condition is the basis of the Gaia Hypothesis. An example is enhanced cloud production resulting from increased emissions of dimethyl sulphide by marine plankton, whose production is

itself increased by rising CO_2 levels. The clouds reflect solar radiation and thus help to counteract the 'greenhouse' warming of the Earth's surface.

Now try the following questions to consolidate your understanding of this Chapter.

QUESTION 7.6 It is often said by marine chemists that weathering consumes acids and releases alkalinity into solution, whereas reverse weathering consumes alkalinity and releases acids into solution. Recalling the definition of alkalinity (Section 6.3.2), can you interpret this somewhat cryptic description?

QUESTION 7.7 If the oceans remain in a steady state and the rate of input of a dissolved constituent increases, what must happen (a) to the rate of removal, (b) to the residence time of that constituent?

QUESTION 7.8 If oceanic waters contain greater concentrations of CO_2 (about 50 parts per thousand by volume) than the atmosphere (about 340 parts per million by volume), how is it that extra atmospheric CO_2 from fossil fuels can be taken up in the oceans?

QUESTION 7.9 When the composition of seawater is corrected for cyclic salts, chloride falls to zero, but sulphate does not (Figure 6.10). Yet sulphur is classified as an excess volatile (Table 6.3), so should not its concentration also fall to zero on correction for cyclic salts? In other words, where does the 'extra' sulphate in river water come from?

APPENDIX CONVERSIONS BETWEEN pH AND $[H^+]$

This Appendix is intended as a reminder about how to manipulate the relationship between pH and the hydrogen ion concentration, $[H^+]$.

By definition

$$pH = -\log_{10}[H^+]$$

There is no difficulty in converting whole number values of pH into $[H^+]$. A pH of 8, for example, means that $[H^+] = 10^{-8}$ mol l^{-1}; a pH of 5 means that $[H^+] = 10^{-5}$ mol l^{-1}.

The problem comes when we wish to convert a pH value of, say, 7.4 into $[H^+]$ in mol l^{-1}.

The first thing is to write $[H^+] = 10^{-7.4}$ mol l^{-1}; this tells you straight away that $[H^+]$ must lie between 10^{-7} and 10^{-8} mol l^{-1}. So you know what to aim for.

The simplest procedure is set out below.

From the definition of pH

$$7.4 = -\log_{10}[H^+]$$

or

$$-7.4 = \log_{10}[H^+]$$

This next step is the most important one:

$$-7.4 = -8 + 0.6$$

The logarithm is now in two parts that can be treated separately. You know from the preamble that the -8 part becomes 10^{-8}, but what about the $+0.6$? The anti-logarithm of 0.6 is 4, to the nearest whole number, so

$$[H^+] = 4 \times 10^{-8} \text{ mol } l^{-1}$$

which is in agreement with the result we anticipated from inspection.

Using exactly the same procedure you can now do another example for yourself.

The pH of a seawater sample is 8.3. What is its hydrogen ion concentration. $[H^+]$, in mol l^{-1}?

By definition

$$8.3 = -\log_{10}[H^+]$$
$$-8.3 = \log_{10}[H^+]$$

By inspection, you can see that $[H^+] = 10^{-8.3}$ mol l^{-1}, so you know it must lie between 10^{-8} and 10^{-9}:

$$-8.3 = -9 + 0.7$$

The -9 part of your 'composite' logarithm becomes 10^{-9} and the anti-logarithm of 0.7 is close to 5, so the hydrogen ion concentration of seawater with pH of 8.3 is

$$[H^+] = 5 \times 10^{-9} \text{ mol } l^{-1}$$

The reverse procedure, the conversion of $[H^+]$ into pH, will of course present few problems when simple integers are involved. When $[H^+]$ is 10^{-6} mol l^{-1}, the pH is obviously 6.

But what about a hydrogen ion concentration of 7×10^{-8} mol l^{-1}?

The first thing to do is to inspect the number and realize that it lies between 10^{-7} and 10^{-8}. So the pH will lie between 7 and 8.

From the definition of pH, you know that you must take the logarithm of 7×10^{-8}, and you know from the nature of logarithms that you need only to *add* the logarithms of these two numbers. The logarithm of 10^{-8} is -8, and the logarithm of 7 is 0.85 to two decimal places. So the logarithm of this value of $[H^+]$ is

$$-8 + 0.85 = -7.15$$

By definition, then, the pH in this case is 7.15. We said that from inspection you would expect it to lie between 7 and 8, and you can see from the size of the number that $[H^+]$ must lie closer to 10^{-7} than 10^{-8}, and similarly the pH is closer to 7 than to 8.

Now try another one. What is the pH of a sample of seawater having a value for $[H^+]$ of 6×10^{-9} mol l^{-1}?

By inspection, that value lies between 10^{-8} and 10^{-9}, and nearer to 10^{-8} than 10^{-9}, so the pH will be between 8 and 9, and probably nearer to 8.

The logarithm of 6×10^{-9} becomes:

$$-9 + 0.78 = -8.22$$

By definition, the pH is therefore 8.22, just as you predicted by inspection.

SUGGESTED FURTHER READING

BOLIN, B., DOOS, B. R., JAGER, J., AND WARRICK, R. A. (eds.) (1987) *The Greenhouse Effect, Climatic Change and Ecosystems*, Scope 29, Wiley. A collection of papers from a major international Conference (Villach, Austria, 1985) convened to discuss many aspects of this important topic.

BROECKER, W. S., AND PENG, T. S. (1983) *Tracers in the Sea*, Eldigio Publications. An advanced text describing the behaviour of many dissolved constituents of seawater in some detail.

DEVOY, R. J. N. (ed.) (1987) *Sea Surface Studies: A Global View*, Croom Helm. An interdisciplinary review of the state of knowledge of sea-level change and its causes, and of the oceanographic, environmental, biological and geological implications.

HENDERSON-SELLARS, A., AND ROBINSON, P. J. (1986) *Contemporary Climatology*, Longman. Deals with long-term climatic changes, climate modelling, and the role of changing CO_2 concentrations in climatic modification.

LISS, P. S., AND CRANE, A. J. (1983) *Man-Made Carbon Dioxide and Climatic Change—A Review of Scientific Problems*, Geo Books, Norwich. More detailed information and insights into this important topic.

LOVELOCK, J. E. (1979) *Gaia: A New Look at Life on Earth*, Oxford University Press. Written by its originator, this is an account of the hypothesis that terrestrial life controls the Earth's surface environment.

STOWE, K. (1983) *Ocean Science* (2nd edn), Wiley. A general text covering all aspects of oceanography, with emphasis on the multi- and interdisciplinary nature of this subject.

ANSWERS AND COMMENTS TO QUESTIONS

CHAPTER 1

Question 1.1 *IMPORTANT:* This answer is much fuller than any we expect you to give. It should raise further questions in your mind, to which you will find the answers as you read on.

The oceans are *salty* because water is a very good solvent, and rivers bring vast amounts of dissolved salts to the sea each year.

The oceans are more or less permanently *cold* below the surface layers (which nowhere extend deeper than a few hundred metres) because they are heated mainly from the top, and water is a poor conductor of heat with a high specific heat. Indeed, the main way in which heat is transferred downwards is not conduction, but turbulent mixing (see Section 2.3). Cold seawater is generally denser than warm seawater, but ice is less dense than liquid water, so in polar regions the oceans are cold from top to bottom (sea-ice forms less readily than ice on a lake because seawater freezes at a lower temperature and cold seawater can sink at all temperatures down to the freezing point—*cf.* Question 4.14.)

The oceans are *dark* below depths of a few hundred metres, because water is a poor transmitter of light; but it transmits sound very well, so the oceans are very *noisy* at certain frequencies.

The oceans *teem with life*, because water is a good solvent and seawater is therefore rich in the nutrients essential to living organisms. Also, there is a constant supply of water, which is an essential and major constituent of all life forms on Earth.

The oceans are *never still* because water is such a mobile liquid. Waves, tides, and currents keep the water in constant motion.

It would also be legitimate to mention the boiling and freezing points of pure water at 100°C and 0°C. Liquid water only exists at the surface of the Earth because surface temperatures lie between 0°C and 100°C nearly everywhere. (This may be a circular argument: water has high specific and latent heats, so the buffering effect of the oceans has helped to keep surface temperatures of the Earth within this range.)

Question 1.2 The 'maximum density' must increase with falling temperature. The greater the dissolved salt content, the more salts are dissolved in the water, and the greater the density.

Question 1.3 (a) Figure 1.3 shows that 336 'units' of water evaporate from the oceans annually (where 1 'unit' = 10^{15} kg). Precipitation in the oceans is 300 'units' per year, and run-off adds another 36 (note that this figure includes the seepage of groundwater). $300+36=336$, so the processes are in balance.

(b) The oceans contain 1322000 'units' and the annual input/output rate is 336 'units', so:

$$\text{residence time} = \frac{1\,322\,000}{336} \approx 3\,900 \text{ years}$$

(c) 336 'units' of water evaporate from the oceans annually, and another 64 'units' evaporate from land. This total of 400 'units' is the amount that moves through the atmosphere annually, for the evaporation is balanced by precipitation and run-off. Note, however, that there are on average only about 13 'units' in the atmosphere at any one time, so the residence time of water in the atmosphere is measured in days.

Question 1.4 (a) The first reason is that liquid water is more dense than ice (Table 1.2). The second reason is that dissolved salts increase the density of water (Figure 1.2).

(b) The dissolved salts have both lowered the freezing point *and* increased the density of the water that forms the droplets. The salt droplets will therefore 'melt' their way through the bottom of the ice. The older the ice, the longer this process will have been operating.

Question 1.5 (a) Both the specific heat and the latent heats of fusion and evaporation are extremely high. Large amounts of heat are required even to raise the temperature of water, let alone to evaporate it. The enormous volume of oceanic waters thus provides a huge temperature buffer, which confines the range of temperatures at the Earth's surface to a few tens of degrees centigrade.

(b) Pure water reaches its maximum density at 4°C, well above the freezing point. The density then decreases again from 4°C to 0°C (Table 1.2). But in seawater, this does not happen; the density increases right down to freezing point (Figure 1.2), which is very important for the control of the deep ocean circulation, as you will see later on.

Question 1.6 The total volume of water on land at any one time approximates to 38000×10^{15}kg (Figure 1.3). The input/output rate (precipitation/evaporation and run-off) for land areas is 100×10^{15}kg per year. So:

$$\text{residence time} = \frac{38 \times 10^{18}}{100 \times 10^{15}} = 380 \text{ years}$$

This average will conceal considerable variations because run-off into streams and rivers is a great deal more rapid than the movement of glaciers or the infiltration and movement of groundwaters. Groundwater and water frozen in ice-caps will both have much greater residence times than the average of 380 years suggests.

Question 1.7 (a) Arctic ice is mainly sea-ice, formed when seawater freezes in the Arctic Ocean basin. Antarctic ice comes mainly from the vast ice-sheet covering the Antarctic continent. The huge ice-shelf round the continent is also augmented by the direct freezing of seawater.

(b) According to Figure 1.2, the sample of water will (i) attain its maximum density at about −0.3°C, and it will (ii) freeze at about −1°C.

(c) Ice melting into seawater will add freshwater and will thus dilute the seawater, making it less saline. This will raise the freezing point of the seawater, and will therefore facilitate the formation of further sea-ice when temperatures fall once more.

CHAPTER 2

Question 2.1 Assuming the albedo of a sandy beach to be more or less comparable with that of desert sands, then a sandy beach will reflect more insolation than a pine wood, i.e. it will *absorb* less solar radiation. If it absorbs less short wave radiation, it will re-emit less long wave radiation back to the atmosphere. So, the atmosphere above a pine wood should be warmed more than that above a sandy beach.

Question 2.2 The three reasons are:

1 Water is transparent, so the radiation penetrates some distance below the surface (but see Figure 2.5), and because water is a liquid, heat is carried to deeper levels by mixing (see Section 2.3).

2 Water has a much higher specific heat (Table 1.1) than rock or soil, and absorbs more heat for a given rise in temperature.

3 Water has very high latent heats of evaporation and fusion (Table 1.1). Large amounts of heat are required to achieve evaporation of water or melting of ice, without *any* rise in temperature.

Question 2.3 (a) From Figure 1.3, about $336 \times 10^{15}/365$ kg of water leave the oceans daily by evaporation, i.e.

$$920 \times 10^{12} \text{kg day}^{-1}$$

So, the heat lost per day is:

$$2.25 \times 10^{6} \text{J kg}^{-1} \times 920 \times 10^{12} \text{kg} = 2.25 \times 920 \times 10^{18} \text{J} \approx 2.1 \times 10^{21} \text{J}$$

(b) 2.1×10^{21} J approaches one-quarter of the daily global insolation value given as 9×10^{21} J, so the loss of heat by evaporation from the oceans is a very important component of the Earth's heat budget.

(c) When warm saturated air passes over a cold sea-surface, water will condense on the surface, giving up its latent heat as it does so. Advection fog would provide evidence of this process.

Question 2.4 (a) Profile I in Figure 2.6(b) corresponds to line A in Figure 2.6(a), and profile II matches with line B. The exact upper and lower limits of the thermocline are most easily defined for the equatorial profile (II), where the thermocline is nearer the surface and has a steeper gradient than the thermocline for the mid-latitude station (profile I).

(b) At high latitudes, there is no thermocline and a temperature–depth profile would show a virtually straight line (see Figure 2.7).

Question 2.5 The curve would probably lie mainly between 18°C and 19°C, and would probably be a more or less horizontal line.

Question 2.6 (a) Table 1.2 shows that freshwater reaches its maximum density at 4°C, so a surface layer of water below this temperature would be less dense than underlying warmer water. This is a gravitationally stable situation, and the less dense water below 4°C would remain at the surface. The presence of dissolved salts in seawater means that the density increases right down to freezing point (Figure 1.2). Water below 4°C will be more dense than underlying warmer layers. That is a gravitationally unstable situation and the colder water at the surface will sink—*unless* the underlying layers happen to be sufficiently more saline to be denser, an aspect we shall explore again in Chapter 4.

(b) The answer again lies in the maximum density of freshwater being at 4°C. As soon as the temperature of the water falls below this, it becomes less dense and it will not sink. Water at more than 4°C is likewise less dense. The densest water in a freshwater lake will always be that which is at 4°C, and for the lake to be gravitationally stable this water must be at the deepest part of the lake.

Question 2.7 In broad terms, as shown in Figure 2.9, we can recognize first the *mixed surface layer*, in which diurnal and seasonal thermoclines occur at certain latitudes and times of year. Below this lies the *main* or *permanent thermocline*, where temperatures fall relatively quickly from more than 15°C to about 5°C. Finally, the bulk of the deep oceans (the *deep layer* from around 1000m to the bottom) is characterized by temperatures of less than about 3°C. This three-layered structure is not seen in high latitudes where temperature profiles are practically vertical (Figure 2.7(c)).

CHAPTER 3

Question 3.1 Table 3.1 gives ionic concentrations by weight. To assess the relative balance of positive and negative ions, we need to divide each ionic concentration figure by the appropriate relative atomic or molecular mass to get its ionic proportion. When we do this for Na^+ and Cl^-, we find that their ionic proportions are much closer than their actual concentrations:

$$Na^+ : \frac{10.6}{23} = 0.46 \qquad Cl^- : \frac{19.0}{35.5} = 0.53$$

You can verify the overall balance of the ions in Table 3.1 by doing the same for all the other positive and negative ions, and adding up the totals.

Question 3.2 Only four of the elements in Table 3.2 appear in Table 3.1: Ca, K, Na and Mg, though not in the same sequence of relative abundance.

Question 3.3 (a) From Table 3.1, the ratio is:

$$\frac{\text{concentration of } K^+}{\text{total salinity}} = \frac{0.380}{34.482} = 0.0110$$

(b) (i) For a salinity of 36:

$$\frac{\text{concentration of } K^+}{36} = 0.0110$$

$$\text{concentration of } K^+ = 0.0110 \times 36 = 0.396 \text{ (‰ by weight)}$$

(ii) and for a salinity of 33:

$$\text{concentration of } K^+ = 0.0110 \times 33 = 0.363 \text{ (‰ by weight)}$$

(c) From Table 3.1, $K^+ : Cl^- = \frac{0.380}{18.980} = 0.0200$

This ratio should be the same in each of cases (i) and (ii) of part (b), and for any other value of salinity in the open oceans, because the relative proportions (the ionic ratios) of major ions remain constant.

(d) Salinity will increase if water is withdrawn from the oceans by evaporation or by the formation of ice; and it will decrease if water is added to the oceans by precipitation, run-off from rivers, or melting ice.

Question 3.4 If oxygen is taken from sulphate (SO_4^{2-}) ions, the sulphate will be reduced to sulphide (S^{2-}). The ratio of SO_4^{2-} to salinity will decrease, with respect to normal seawater.

Question 3.5 (a) Profile I on Figure 3.3(a) corresponds to line A on Figure 3.3(b). Profile II corresponds to line B. The depth range of the halocline on each profile corresponds quite closely with that of the thermocline at the same location on Figure 2.6.

(b) The halocline is nearer the surface for the equatorial profile (II), and the decrease of salinity with depth across the halocline is much sharper—you can tell that from Figure 3.3(a), because the isohalines are crowded together in the equatorial belt.

(c) No. At *high* latitudes, low salinity values are found from the surface to the bottom, so the profiles would approximate to vertical lines. There is no halocline there.

Question 3.6 Surface salinities are controlled by the balance between evaporation and precipitation. Figure 3.4(b) shows how close this relationship is: where precipitation exceeds evaporation, the salinity is low; where evaporation is high, so is the salinity. The salinity minimum (see also Figure 3.3(a)) is due to an excess of rainfall over evaporation along the equatorial belt, despite high average insolation and temperatures.

Question 3.7 35. If K_{15} is equal to 1, equation 3.4 simplifies as follows:

$$S = 0.0080 - 0.1692 + 25.3851 + 14.0941 - 7.0261 + 2.7081 = 35.$$

Question 3.8 (a) False. Tables 3.1 and 3.2 show that the proportions are quite different.

(b) True. Chloride is a major constituent of seawater and its ratio to salinity is constant throughout most of the oceans. That is why chlorinity measurements were for a long time the principal means of determining salinity.

(a) True. If Ca^{2+} ions are selectively removed from solution, then the Ca:S ratio must fall.

(d) False, at least in general. Typically, the halocline represents a decrease of salinity with depth, but in some cases 'reverse' haloclines can develop.

(e) False. However, in terms of practicalities, routine measurements of both salinity and temperature have a precision of about one part in a thousand (Section 3.3.2).

Question 3.9 If dissolved salts are being continually added to seawater and it nevertheless maintains a constant composition—both in the total amount and the relative proportions of dissolved constituents—then elements must be removed (precipitated) from seawater (by biological activity or chemical precipitation), *at the same rates* at which they are being added. If this were not the case, there would be much greater variations of salinity throughout the oceans and ratios of major elements both to one another and to total salinity would fluctuate a lot more than they do. We shall be exploring this in Chapter 6.

CHAPTER 4

Question 4.1 (a) The boundaries are sketched in on Figure A1. The three water masses are known as: 1, Antarctic Bottom Water (AABW); 2, North Atlantic Deep Water (NADW); 3, Antarctic Intermediate Water (AAIW). See also Section 4.3.3.

(b) (i) Within individual water masses, temperature, salinity and other properties are fairly uniform throughout. (ii) At boundaries between water masses, however, there can be quite sharp changes. The gradients of change in T and S across boundaries gradually become less abrupt as water masses move further from their source regions (Figure A1) and sufficient time has elapsed for the boundaries to become blurred by mixing.

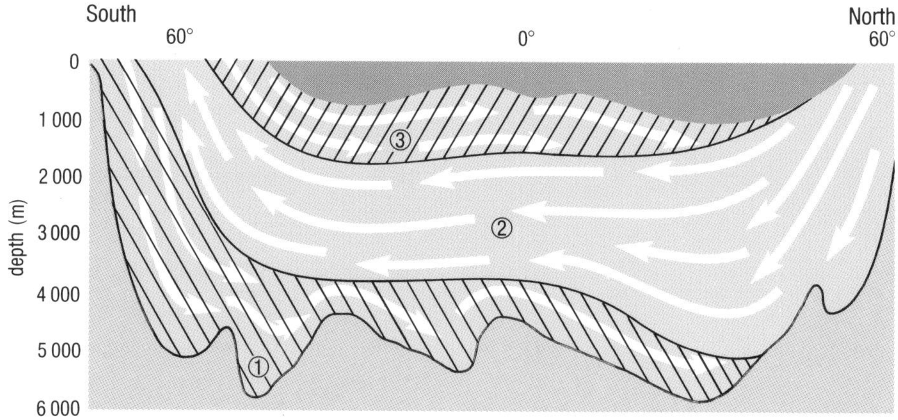

Figure A1 Answer to Question 4.1.

Question 4.2 (a) (i) Surface waters in polar regions are cooled both by direct contact with ice-sheets and by cold winds blowing off the ice. Ice formed when seawater freezes leaves behind seawater of slightly higher salinity. Both these processes combine to increase the density of surface waters (*cf*. Figure 1.2).

(ii) In tropical regions, the salinity (and therefore the density) of surface waters is increased mainly by evaporation (*cf*. Figure 3.4). Occasional strong winds can also cool the surface waters, counteracting the effect of high insolation which acts to warm the water and lower the density.

(b) Yes, of course: cold air at the poles extracts thermal energy from the surface of the ocean, either directly by blowing over it, or indirectly through contact with ice, itself a product of precipitation from the atmosphere. The cold, dense water so formed sinks to the ocean floor. In lower latitudes, heating of the ocean surface ensures that the water remains generally less dense there. This heating contributes substantially to the total global hydrological cycle by causing evaporation. Some of the water evaporated in low latitudes is transported in the atmosphere towards polar regions, where the water falls as snow.

Question 4.3 (a) Substituting in the hydrostatic equation (equation 4.1):

$$\begin{aligned} \text{pressure} &= 1.03\times10^{3}\,\text{kg}\,\text{m}^{-3}\times9.8\,\text{m}\,\text{s}^{-2}\times10\,\text{m} \\ &= 10^{5}\,(\text{kg}\,\text{m}\,\text{s}^{-2})\,\text{m}^{-2} \\ &= 10^{5}\,\text{N}\,\text{m}^{-2} \end{aligned}$$

According to the caption to Figure 4.3, the pressure due to a 10m column of seawater is close to normal atmospheric pressure, which can also be expressed as 1000 millibars (mbar).

(b) (i) The deep ocean floors are around 4–5km deep, which corresponds to pressures in the order of 4×10^{7} to $5\times10^{7}\,\text{N}\,\text{m}^{-2}$ or 400–500 atmospheres.

(ii) Some ocean trenches reach depths of about 10km, so pressures there must be about $10^{8}\,\text{N}\,\text{m}^{-2}$, or 1000 atmospheres.

Question 4.4 (a) Air would be subjected to *increased* pressure in the adiabatic change from a height of 5km down to sea-level. It would become more compressed, so its potential temperature will be *higher* than its *in situ* temperature.

(b) Seawater would be subjected to *reduced* pressure in the adiabatic change from a depth of 5km up to sea-level. It would become less compressed, so its potential temperature will be *lower* than its *in situ* temperature.

Question 4.5 (a) For (i), σ_t is 27.6 (interpolating between the contours on Figure 4.4); for (ii), σ_t is 26.4.

(b) These two σ_t values 'translate' into (i) specific gravity values of 1.0276 and 1.0264 respectively, and into (ii) density values of $1.0276\times10^{3}\,\text{kg}\,\text{m}^{-3}$ and $1.0264\times10^{3}\,\text{kg}\,\text{m}^{-3}$, respectively. (Specific gravity of a substance is equal to:

$$\frac{\text{density of substance}}{\text{density of pure water}}$$

and density of pure water $= 1.000\times10^{3}\,\text{kg}\,\text{m}^{-3}$.)

(c) The range of temperature in the oceans is of the order of 0–25°C, whereas that of salinity is in general little more than about 34.5–35.5. Temperature typically has a greater effect on density than salinity. In addition, the contours on Figure 4.4 mostly make a smaller angle with the salinity axis than with the temperature axis. For temperatures greater than about 5°C, a change in temperature of 1°C has a greater effect on density than a change in salinity of 0.1.

Question 4.6 (a) Figure 4.6(a) must correspond to description (i) because it has the warmer bottom water; Figure 4.6(b) matches description (ii) as it shows cool water at the bottom of the trench.

(b) The increase of temperature below 4000m in Figure 4.6(a) shows that this must represent the *in situ* temperature part of Table 4.1. Figure

4.6(b) must therefore represent the potential temperature part. Potential temperature profiles and sections are more realistic because corrections have been made for adiabatic compression. So, description (ii) is correct.

Question 4.7 (a) Most of the curve crosses the contours of equal density in such a way that it shows a consistent increase of σ_t with depth, and suggests that the water column is stable down to at least 2000m. Below that, the line is roughly parallel with the contours, suggesting a more or less constant density.

(b) σ_t can only give a rough indication of stability, for it is uncorrected for adiabatic heating (by definition), and so it gives spurious information about density (and hence stability), especially in deep water. A more reliable indication of stability would be obtained by plotting σ_θ against depth.

(c) The curve would be similar in shape but would be displaced progressively downwards with respect to the curve in Figure 4.7, with increasing depth. The point at 150m depth would be virtually the same for both profiles, and the divergence would increase with depth, becoming greatest at 5000m, where Table 4.1 suggests that the potential temperature would be *c.* 0°C.

Question 4.8 (a) Ca^{2+}, HCO_3^- and Mg^{2+} participate in processes that 'cycle' them within the seawater solution. Their concentrations are changed by processes other than mixing. They therefore behave non-conservatively.

(b) Within the main body of the oceans, chloride, iodide and bromide do not participate in any process that removes them from the seawater solution. Their concentrations are changed only by mixing, and so chlorinity is a conservative property.

Question 4.9 The density of seawater is increased both by direct cooling and by the removal of freshwater to form ice, which increases the salinity of the seawater left behind. It becomes gravitationally unstable and sinks, and then moves towards equatorial regions. The main sources of these cold water masses are in the Greenland, Norwegian and Labrador Seas and in the Antarctic Weddell Sea.

Question 4.10 No. In both cases the temperature falls with increasing distance from the Earth's surface, where most of the insolation is absorbed (directly or indirectly). The resemblance ends there. Decrease of temperature with depth in the oceans is due partly to the limited penetration of sunlight (see also Chapter 5); and partly to the limited downward extent of vertical mixing. Temperatures at depth are also low because of the thermohaline circulation, which supplies cold water from polar regions to the deep oceans. Decrease of temperature with height in the troposphere is due mainly to the adiabatic expansion of air as it rises on being warmed at the Earth's surface.

Question 4.11 Specific gravity values of 1.025 and 1.026 are represented by σ_t values of 25.0 and 26.0. From Figure 4.4, if σ_t is to increase by the amount indicated, then either: (a) the temperature must decrease by about 5°C; or (b) the salinity must increase by about 1.3.

(c) The definition of σ_t means that density is calculated as if the water sample were under atmospheric pressure. Water at a depth of 4000m will be under a pressure of some 400 atmospheres (Figure 4.3). So its *in situ* temperature will be higher than its potential temperature, and the density corresponding to its σ_t value of 27.6 is bound to be less than its true density.

Question 4.12 (a) The large spread of values must represent mainly near-surface waters. Examination of any profiles (e.g. Figure 2.6, Figure 3.3) shows that the largest ranges of *T* and *S* are in the upper 500m of the water column.

(b) The three values all plot together in the bottom of Figure 4.16 but the small spread corresponds to that of the main fields shown on the diagram (i.e. Pacific on the left, Atlantic on the right, Indian in the middle). The *average* values for ocean waters will be determined largely by the great mass of water of relatively low temperature and salinity below the main halocline and the permanent thermocline.

Question 4.13 (a) False. Salt fingering involves molecular diffusion; internal wave breaking is a process of turbulent mixing.

(b) True. *In situ* temperature is subject to change not only by mixing, but also by adiabatic heating or cooling as a result of changes in pressure. Hence it does not fall within the strict definition of conservative properties. Potential temperature has been corrected for adiabatic changes, and is therefore a true conservative property.

(c) False. Only the *in situ* temperature can be measured. Potential temperature is obtained by correcting the measured temperature for the effects of adiabatic compression.

(d) True. Both processes lower the density of surface waters: warm water is less dense than cold water, and freshwater is less dense than seawater.

Question 4.14 (a) By extrapolation from Figure 4.4, (i) water at −1°C and with salinity of 33 has σ_t of about 26.7; while (ii) water at −1°C and with salinity of 35 has σ_t of 28.2. These correspond to densities of $1.0267 \times 10^3 \mathrm{kg\,m^{-3}}$ and $1.0282 \times 10^3 \mathrm{kg\,m^{-3}}$.

(b) Water of salinity 33 would freeze at about −1.8°C, from Figure 1.2, but even as it approached this lower temperature its density would not increase above that of the underlying water of salinity 35. Sea-ice could therefore form.

CHAPTER 5

Question 5.1 (a) (i) No. Even at the surface of the ocean, moonlight has an intensity about four orders of magnitude lower than that required for phytoplankton growth. (ii) No, normally the limiting depth in even the clearest coastal water in sunlight is about 50m. (iii) Yes. Indeed, phytoplankton can grow *only* during daylight within this narrow depth zone, and if the water is not clear then the range is further reduced.

(b) (i) No. Moonlight of the required minimum intensity for perception penetrates only to a depth of about 700m. (ii) Yes, but this approaches the limit: below about 1250m, they must live in virtually total darkness.

Question 5.2 Dark coloration will reduce contrast when the fish is viewed from above or from the side, i.e. against a (normally) dark bottom and dark surrounding water. A silvery underside will similarly reduce the contrast when the fish is viewed from below, against surface waters illuminated by daylight, i.e. against a background of downwelling irradiance. In both cases, the fish is less obvious to both predator and prey.

Question 5.3 (a) The coefficient of attenuation for directional light must always be greater than the coefficient of diffuse attenuation for non-directional light. That is because an underwater surface can be illuminated by light from any direction, even though it may be appreciably scattered; whereas an object can only be perceived as a result of light that has travelled in a direct line from the object to the eye (or camera)—and most of the light leaving the object along the required path will be scattered away from that path.

(b) Sunlight penetrates furthest through the clearest ocean water, so this must have the smallest coefficients of attenuation. Coastal water is generally more turbid and so has larger coefficients.

Question 5.4 (a) From equation 5.2:

$$K \times 10 = 1.5$$
$$K = 0.15$$

As the product of $K \times Z_s$ can range from 1.4 to 1.7, the potential error in K can be as much as 2/15, or of the order of 10–15%.

(b) (i) From equation 5.3:

$$V = 0.7 \times 20 = 14\text{m}$$

(ii) From equation 5.4:

$$Z_e = 3 \times 20 = 60\text{m}$$

You get the same result, of course, using equation 5.5 and the answer from (i).

Question 5.5 For horizontal vision, the sighting angle θ is zero, $\sin\theta$ is zero, and so is the term $K \sin\theta$ in equation 5.6. So, horizontal visibility is independent of K.

Question 5.6 Blue–green light (i.e. shorter wavelengths in the visible spectrum) is preferentially absorbed by the photosynthesizing algae, for conversion to chemical energy. Blue–green wavelengths are preferentially transmitted by clear seawater (Figures 2.5, 5.7): the less turbid the water, the greater the depth at which photosynthesis can proceed (the deeper the photic zone). In turbid coastal waters, high concentrations of suspended matter and yellow substances result in absorption of blue–green wavelengths (*cf.* Figure 5.7), as well as greater scattering of the incident light (higher values of K). The depth of the photic zone is reduced and photosynthetic activity tends to be inhibited.

Question 5.7 The answer is *not* $3 \times 35 = 105\%$. If 35% is absorbed, then 65% is transmitted, and this becomes the 'incident light' for the next 1m, and so on. So, in 3m, the transmission will be:

$$65\% \text{ of } 65\% \text{ of } 65\% \text{ or } 65 \times 0.65 \times 0.65 = 27.5\%.$$

That is the proportion transmitted, so the proportion absorbed is $100-27.5=72.5\%$.

Question 5.8 The speed of sound in seawater is less than that in rock, but it is greater than that in air (see also Table 5.1). The reason for the apparent anomaly is that, in general, denser materials have higher axial moduli—they are 'stiffer' (i.e. less compressible). The axial modulus is in the numerator of equation 5.8, and in general the axial modulus increases more than density.

Question 5.9 Substituting in equation 5.9:

$$c=1410+(4.21\times10)-(0.037\times10^2)+(1.14\times35)+(0.018\times100)$$
$$=1490.1\,\text{m}\,\text{s}^{-1}.$$

Question 5.10 If $Z_1=Z_2$, then Z_1-Z_2 is zero, and R must also be zero. Sound will cross the interface with little or no loss of acoustic energy.

Question 5.11 (a) The speed of sound in the sound channel changes significantly from one end of the section to the other. It is less than $1470\,\text{m}\,\text{s}^{-1}$ at 50°S, reaches nearly $1500\,\text{m}\,\text{s}^{-1}$ at 30°N and then declines again, down to about $1480\,\text{m}\,\text{s}^{-1}$.

(b) The depth of the sound channel axis is greatest at around 30°N, remains at around 1km between about 10°N and 40°S, and becomes very shallow above about 50° latitude. The main reason for the fluctuations is that there are variations in the depth of the thermocline (*cf.* Figure 2.9); at higher latitudes the whole water column is well mixed, so the permanent thermocline and main halocline are absent (*cf.* Figure 2.7(c)).

(c) From Section 5.2.2, increases in both T and S lead to an increase in the speed of sound, and pressure has a relatively small effect in the surface layers. At depth, neither T nor S change much, but pressure increases the axial modulus of the water more than it increases the density. So, from equation 5.8, c must increase.

Question 5.12 (a) A change in temperature of 1°C leads to a change in sound speed of the order of $3\,\text{m}\,\text{s}^{-1}$, whereas a change in salinity of 1 changes the speed of sound by only $1.1\,\text{m}\,\text{s}^{-1}$ (Section 5.2.2). A change in salinity of 0.1 thus affects the speed of sound hardly at all in comparison with the temperature change of a degree or two.

(b) If you look back to Figure 5.10 or 5.14, you will see that speeds in the sound channel are at a minimum, and increase to a maximum at the sea-bed. Even though sound travelling down to the bottom and back may travel further, it still travels faster and arrives earlier than sound in the sound channel, because most of its path is in a medium that transmits sound at a higher speed.

Question 5.13 Because they have been absorbed in the topmost 10m of the water column, *cf.* Figure 2.5.

Question 5.14 Some fish have an air-filled swim-bladder which occupies only about 5% of the total volume of the fish, yet may account for more than 50% of the returning echo, because of the very high reflectivity of air–water interfaces (Table 5.1). Indeed, air bubbles flowing along the hull of a heaving fishing boat can blanket sonar equipment and render the fish-finder device almost inoperable. (Cartilaginous fish such as sharks, rays, skates and dogfish, do not have swim-bladders; nor do some boney fishes, such as mackerel and tuna. Swim-bladders of fishes that live below about 1000m are filled with fat, not gas, and swim-bladders of sea-bed dwelling fish are used as auditory organs rather than for buoyancy, Section 5.2.3.)

Question 5.15 (a) To detect individual fish in a shoal or near the sea-bed requires good range discrimination; so, fishing sonars are high frequency, short range systems.

(b) Presumably, one would wish to detect a submarine at maximum possible range; however, the size of such a target would not require highly accurate range discrimination. Antisubmarine sonars therefore operate at lower frequencies to maximize the range. This also means that they must employ large and expensive transducer equipment to maintain a narrow beam for bearing discrimination.

Question 5.16 (a) False. Open ocean water is clearer and more blue than near-shore water, which is often yellowish as well as turbid. Light penetration in coastal water is less and is biased away from the blue–green light required by most photosynthesizing organisms.

(b) False. The water is described as being of an 'intense blue', which means it is non-productive biologically (Section 5.1.4).

(c) True. Water below the thermocline is usually at less than 6°C, which is the lower limit of temperature given for equation 5.9.

(d) True, but does it matter? The point is that the acoustic impedance of seawater is so enormous compared with that of air (Table 5.1)—nearly four orders of magnitude greater—that for all practical purposes, the acoustic reflectivity of the air–sea interface is 100%.

Question 5.17 The broken line represents the mean winter profile of sound speed with depth. As temperature falls, the speed of sound decreases. The depth of the sound channel axis is about 600–700m, and changes at and below this depth are minimal because seasonal changes penetrate no more than a few hundred metres below the surface (*cf*. Figure 2.8).

CHAPTER 6

Question 6.1 Table 6.1 gives the concentrations of individual elements, whereas in Table 3.1 the concentrations are listed in terms of predominant ionic *species*, i.e. the form in which the dissolved constituents occur (see Section 6.3.1). For the simple ions, e.g. Na^+ and Cl^-, ionic and elemental concentrations are the same, but for most anions they are clearly different. Sulphur occurs predominantly as the sulphate anion (SO_4^{2-}). Carbon is present as carbon dioxide (CO_2), carbonic acid (H_2CO_3) and its dissociation products (HCO_3^- and CO_3^{2-})—at the pH of seawater (8.0 ± 0.2) HCO_3^- is the dominant form—together with a small but rather variable amount (about $1\,mg\,l^{-1}$) in the form of dissolved organic molecules. Boron occurs as the hydroxide ($B(OH)_3$) and related ionic forms. The disparities can be avoided by presenting analyses in molar terms. A mole of bicarbonate ion (HCO_3^-) contains a mole of carbon (C), for example, so the concentrations become the same. If that is not clear to you at this stage, do not worry—it should become so later on.

Question 6.2 For this calculation:

$$g = 9.8\,\mathrm{m\,s^{-2}}$$
$$d = 2 \times 10^{-6}\,\mathrm{m}$$
$$\rho_1 = 1.5 \times 10^3\,\mathrm{kg\,m^{-3}}$$
$$\rho_2 = 1.0 \times 10^3\,\mathrm{kg\,m^{-3}} \quad \text{(close enough to seawater for our purposes)}$$
$$\mu = 10^{-3}\,\mathrm{N\,s\,m^{-2}} \quad \text{(close enough to seawater for our purposes)}$$

Then, substituting in equation 6.1,

$$v = \frac{1}{18} \times 9.8 \times \frac{(1.5-1) \times 10^3}{10^{-3}} \times (2 \times 10^{-6})^2$$
$$= \frac{9.8}{18} \times 0.5 \times 10^6 \times 4 \times 10^{-12}$$
$$= \frac{9.8 \times 2}{18} \times 10^{-6} = 1.09 \times 10^{-6}\,\mathrm{m\,s^{-1}}$$

This particle would therefore take about 0.92×10^6 seconds (about 255 hours or more than 10 days) to sink through 1 m.

Question 6.3 Well-stratified waters are gravitationally stable. Nutrients which sink from the surface cannot be replaced except by slow diffusion processes. If the water column is well mixed, on the other hand, nutrients sinking from the surface have a better chance of being carried back up again. A well-developed pycnocline, forming a barrier to upward mixing, will be found in well-stratified waters.

Question 6.4 (a) From Figure 6.4, the ratio of N_2:O_2 in air is close to 8:2 or 4:1, whereas in seawater in equilibrium with the atmosphere it drops to about 9:5.3 or 1.7:1. So the solubility of oxygen is more than double that of nitrogen (4/1.7).

(b) Similarly, the ratio Ar:CO_2 in air is about 30:1, whereas in seawater it is only about 1:250. So, CO_2 is about: $30 \times 250 = 7500$ times more soluble than argon in seawater.

Question 6.5 (a) Chiefly because the source regions are very cold, and Figure 6.5 shows the solubility of oxygen to increase with falling temperature.

(b) The oxygen minimum is most strongly developed in profile I: water at intermediate depths is almost anoxic. This could be due partly to the respiratory needs of large populations of animals supported by high levels of primary production in surface waters, but perhaps mainly to bacterial activity associated with the decomposition of organic detritus sinking out of the photic zone.

Question 6.6 The sea→air flux of N_2O from Table 6.2 is $1.2 \times 10^{14} g yr^{-1}$. That is:

$$1.2 \times 10^{14} \times \frac{28}{28+16}$$

$$= 1.2 \times 10^{14} \times 0.64 \text{ g N yr}^{-1}$$

$$= 7.7 \times 10^{13} \text{ g N yr}^{-1}$$

The balance to be made up is:

$$8.0 \times 10^{13} \text{ g N yr}^{-1} - 0.9 \times 10^{13} \text{ g N yr}^{-1} = 7.1 \times 10^{13} \text{ g N yr}^{-1}$$

The sea→air flux of N_2O gas more than makes up this balance. In fact, the agreement is extremely good, given that global estimates of fluxes of this nature are subject to considerable uncertainties.

Question 6.7 (a) Oxygen is non-conservative, because its concentration is changed by biological or chemical reactions, i.e. not by mixing only. The same applies to most other dissolved gases. The concentration of dissolved oxygen in subsurface water masses decreases with time and with distance from the source regions, because it is used up in biological respiration; and below the photic zone there is no mechanism of replenishment other than downward mixing from the surface, which is usually too slow to keep pace with the rate of consumption.

(b) Figure 6.7(a) shows a high content of dissolved oxygen in surface waters in both polar regions in the Atlantic Ocean. This diminishes gradually with depth and distance towards the Equator, consistent with the sinking of water masses in polar regions. If you compare Figure 6.7(a) with Figure A1, you should be able to see the long 'tongue' of North Atlantic Deep Water, extending southwards at mid-depths. The overlying Antarctic Intermediate Water is also visible. Less obvious is the gradual decrease in dissolved oxygen content from surface to bottom in Antarctic regions that corresponds to sinking Antarctic Bottom Water.

(c) Figure 6.7(b) also shows relatively high concentrations of dissolved oxygen in surface waters in the South Pacific. These diminish northwards and with depth, consistent with the sinking of Antarctic waters. However, in the North Pacific there is no sinking of surface waters rich in oxygen; instead there is a strong oxygen-minimum layer at *c*. 500–1000m depth throughout most of the North Pacific. The dissolved oxygen contours thus give no indication of a source region of deep water in the North Pacific itself, but they do suggest that there is some influx of deep water extending northwards from the Antarctic. The North Pacific is effectively isolated from any possible deep water source in the Arctic Ocean by the shallow barrier of the Aleutian Islands Chain.

Question 6.8 (a) (i) Rainwater is about 5000 times more dilute than seawater $(34.4/(7.1 \times 10^{-3}))$; (ii) river water is about 300 times more dilute than seawater $(34.4/(118.1 \times 10^{-3}))$.

(b) The similarity in composition between rainwater and seawater is particularly striking. Most of the dissolved salts in rainwater have a marine origin, as a consequence of the injection of seawater into the atmosphere as aerosols produced by bubble-bursting at the sea-surface (Section 2.2.1). River water is very different in composition, with relatively much more Ca^{2+} and HCO_3^- and SiO_2, and much less Na^+ and Cl^-.

Question 6.9 Clearly (i) and (iv) must apply to seawater, and (ii) and (iii) to river water corrected for cyclic salts. Note the complete disappearance of Cl^- in Figure 6.10 as a result of this correction.

Question 6.10 The amount of average crustal rock that must be weathered to provide 11 g of sodium in solution is:

$$\frac{11}{1.8} \times 100 \approx 600\,\text{g}$$

Question 6.11 (a) Sodium, potassium, magnesium and chloride have residence times that are changed significantly by making the correction for cyclic salts.

(b) If $2.5 \times 10^8\,\text{t}\,\text{yr}^{-1}$ are added to the $4.88 \times 10^8\,\text{t}\,\text{yr}^{-1}$ in Table 6.4, the uncorrected residence time for calcium becomes significantly shorter.

$$\frac{\text{mass of Ca in oceans}}{\text{rate of supply of Ca to oceans}} = \text{residence time}$$

$$\frac{6 \times 10^{14}}{7.38 \times 10^8} \approx 800\,000 \text{ years}$$

The same would apply to the residence time corrected for cyclic salts.

Question 6.12 (a) The stirring (or mixing) time is the period for an average water molecule to travel from the surface to the deep ocean and back again (about 500 years). The residence time is the period (about 4000 years) spent by an average water molecule actually within the oceans, before being returned to the atmosphere in the hydrological cycle (Figure 1.3).

(b) Not many elements in Figure 6.11 have residence times of 500 years or less. However, iron and aluminium appear to be in the seawater solution for too short a time to permit complete mixing throughout the whole ocean. Other elements with residence times of less than about 10^3 years may also not be completely mixed.

Question 6.13 Mg^{2+} will have the largest hydration sphere relative to its size, as it has the greatest charge density of the three (smallest radius, biggest charge). Chloride will have the smallest hydration sphere relative to its size, as it has the lowest charge density of the three (largest radius, single charge).

Question 6.14 (a) Cl^- is missing from Table 6.5 because it does not form an ion pair with any major cation. In fact, equilibrium constants for interactions between major cations and Cl^- (and, incidentally, between K^+ and HCO_3^- and CO_3^{2-}—see Table 6.5) show that association is not significant in these cases. In other words, these ions are subject only to electrostatic attraction and repulsion, and behave effectively as free ions with respect to one another, as in Figure 6.13(a).

(b) Chloride is the most abundant ion in seawater, and as it does not form ion pairs, this 'burden' falls on the remaining anions, which are present in much lower concentrations. Relatively greater amounts of these will therefore be involved in ion-pair formation with the plentiful cations.

(c) $MgSO_4$ appears to be the most abundant. You encountered it in Chapter 5 (Figure 5.9), in the context of the attenuation of acoustic energy in the oceans.

Question 6.15 Substituting in equation 6.12:

For surface sample: $[CO_3^{2-}] = 2.35 - 2.0 = 0.35\,mol\,m^{-3}$.
For deep sample: $[CO_3^{2-}] = 2.55 - 2.4 = 0.15\,mol\,m^{-3}$.

Substituting in equation 6.10:

For surface sample: $2.0 = [HCO_3^-] + 0.35$, so $[HCO_3^-] = 1.65\,mol\,m^{-3}$.
For deep sample: $2.4 = [HCO_3^-] + 0.15$, so $[HCO_3^-] = 2.25\,mol\,m^{-3}$.

Substituting in equation 6.16:

For surface sample:

$$[H^+] = 1.0 \times 10^{-9} \times \frac{1.65}{0.35} = 4.7 \times 10^{-9}\ mol\ m^{-3}.$$

For deep sample:

$$[H^+] = 1.0 \times 10^{-9} \times \frac{2.25}{0.15} = 1.5 \times 10^{-8}\ mol\ m^{-3}.$$

From equation 6.13 and the Appendix, $pH = -\log_{10}[H^+]$, so:

For surface sample: $\log 4.7 \times 10^{-9} = -9 + 0.7 = -8.3$, and $pH = 8.3$.
For deep sample: $\log 1.5 \times 10^{-8} = -8 + 0.2 = -7.8$, and $pH = 7.8$.

This example illustrates the general case: deep water in the ocean is generally more acid (lower pH) than surface water.

Question 6.16 (a) Seawater is normally an oxidizing medium. We know that from diagrams such as Figures 6.5 to 6.7, which show that there is free oxygen in the water. More generally, we know that the oceans are full of living organisms that need oxygen for respiration.

(b) As normal seawater is an oxidizing medium, iron must be in the trivalent (Fe^{3+}) state, which is the less soluble form.

Question 6.17 (a) The straight line on Figure 6.15 can be extrapolated through the origin. The inference is that when nitrate is exhausted by biological activity, so is copper. Therefore, in these Antarctic waters at least, it would appear that copper is biolimiting. In most parts of the ocean, however, copper is identified as exhibiting bio-intermediate behaviour. The term micro-nutrient is used where the element concerned is a minor or trace constituent in seawater.

(b) The nickel profiles in Figure 6.16 have similarities to those for both phosphate and silica. We could therefore infer that biological processes influence nickel concentrations.

Question 6.18 (a) Nitrate is almost totally depleted in surface water, and is therefore quite obviously a biolimiting constituent.

(b) Barium shows partial depletion in surface water with respect to deep water and is thus a bio-intermediate constituent.

(c) Sodium maintains a constant ratio of concentration to total salinity throughout the depth range and is therefore a bio-unlimited constituent.

Question 6.19 (a) The concentration of manganese in seawater is about 2×10^{-4} p.p.m., according to Table 6.1. At 10 p.p.m., it is $10/(2 \times 10^{-4})$, i.e. 5×10^4 times more concentrated in hydrothermal solutions.

(b) 5×10^{11} tonnes of heated seawater contain $5 \times 10^{11} \times 10 \times 10^{-6}$ tonnes of manganese. That is 5×10^6 tonnes of manganese. The hydrothermal flux is thus able to account almost entirely for the shortfall in the manganese budget.

(c) The annual flux of manganese through the oceans (from source to sink) is of the order of 5×10^6 tonnes. Table 6.1 gives the total mass of manganese in the oceans as 2.64×10^8 tonnes. Its residence time is therefore:

$$\frac{2.64 \times 10^8}{5 \times 10^6} \approx 50 \text{ years}$$

This is very much shorter than the residence time in both Table 6.4 (14000 years) and Figure 6.11 (1000 years), but closer to the latter. The reason is that the entry in Table 6.4 is based only on river flux, whereas that in Figure 6.11 is based on more accurate estimates of input and removal rates than we have used here.

Question 6.20 Anoxic conditions are reducing conditions, so manganese should exist in its lower valency state, i.e. as Mn^{2+} ions in solution.

Question 6.21 (a) False. Only a very small proportion of the nitrogen in seawater (about 0.5 p.p.m., out of a total of 11.5 p.p.m.) is in the form of nitrate (Section 6.1.2).

(b) False for most dissolved gases, because they are involved in chemical and/or biological processes that change their concentrations. Some inert gases could behave conservatively.

(c) True. The (oxygen) compensation depth is where there is a balance between consumption of oxygen by respiration of plants and production of oxygen by photosynthesis. The oxygen minimum layer is where oxygen used in respiration is not replaced by photosynthesis, because it is well below the photic zone.

(d) True. The shorter the residence time, the greater the annual supply from rivers relative to the total mass in the oceans.

(e) False. Ca^{2+} has the greater charge density and will therefore have the relatively larger hydration sphere.

(f) True. Salts will only be precipitated when they reach saturation. Figure 3.1 shows that large amounts of water must be evaporated to make the seawater solution more concentrated, so that saturation is reached (note that $CaCO_3$ is the exception to this, Section 6.3.2).

(g) True. Hydrogen ions are being added to seawater, and from equations 6.7 and 6.8 the result must be to expel CO_2 gas from solution. In short, reaction 6.2 moves to the left.

CHAPTER 7

Question 7.1 The CO_2 in calcium carbonate accumulated in deep-sea sediments and precipitated in oceanic crust during hydrothermal circulation will eventually be recycled by subduction, re-melting and volcanism. Carbon in organic matter preserved in sediments is in reduced form. Heating during subduction converts the organic matter to hydrocarbons, especially methane gas.

Question 7.2 Deep ocean circulation would be relatively sluggish, because large amounts of cold dense water would not be forming in polar regions. Warm water contains less dissolved gas than cold water, so oxygen concentrations should be lower when average temperatures are high. Deep water would therefore be relatively oxygen-poor.

Question 7.3 (a) The present-day partial pressure of CO_2 is 0.0003, from Figure 6.4. On the early Earth it would have been 1000×0.0003 or 0.3 which approximates to the present-day partial pressure of oxygen. In volumetric terms, CO_2 was second only to nitrogen in the atmosphere of the early Earth.

(b) Greater partial pressure of CO_2 in the atmosphere would drive reaction 6.2 to the right; but there is a limit to the amount of carbon dioxide that seawater could absorb above present concentrations, because present-day surface waters are supersaturated with respect to calcium carbonate. All the same, we can expect $[\Sigma CO_2]$ to have been higher in the primitive ocean despite the greater temperatures, so from equations 6.6 to 6.10, $[HCO_3^-]/[CO_3^{2-}]$ should have been higher too, and so should $[H^+]$; hence lower pH and more acid oceans.

Question 7.4 Figure 7.3 shows that maximum insolation occurs: (a) in northern and southern mid-latitudes during the respective summer solstices (Figure 2.2), when days are long and daily values reach more than $25 \times 10^6 J m^{-2}$; (b) along the Equator, where the average insolation is greater than $20 \times 10^6 J M^{-2} day^{-1}$ (the average at mid-latitudes over the year is little more than $15 \times 10^6 J m^{-2} day^{-1}$).

Question 7.5 The present-day concentration is nearly 345 p.p.m. by volume (Figure 7.4), compared with less than 300 p.p.m. 130000 years ago (Figure 7.2). More remarkable, the increase of 55 p.p.m. from the year 1800 to the present has occurred in less than 200 years. At the start of the last deglaciation, a comparable rise took 1000 years or more. The rate of increase of atmospheric CO_2 is probably greater now than it has been at any time in the Earth's history.

Question 7.6 Alkalinity is represented by carbonate and bicarbonate. Hydrogen ions (acid) in rainwater are neutralized (consumed) during rock weathering, and bicarbonate and carbonate ions (alkalinity) are released into solution. In reverse-weathering reactions, the bicarbonate and carbonate ions (alkalinity) are removed from solution into solid phases (consumed), releasing hydrogen ions (acid) back into solution. A simple example is provided by reaction 7.1 which moves to the right for weathering and to the left for reverse weathering.

$$\underset{\text{(hydrogen ions in solution— acid)}}{H^+} + \underset{\text{calcium carbonate}}{CaCO_3} \underset{\text{reverse weathering}}{\overset{\text{weathering}}{\rightleftharpoons}} Ca^{2+} + \underset{\text{(bicarbonate ions in solution – alkalinity)}}{HCO_3^-} \quad (7.1)$$

Question 7.7 In a steady-state ocean, if the rate of input of a constituent increases, then (a) the rate of removal must also increase, and (b) the residence time must decrease, because the concentration and hence the total mass in the ocean should not change.

Question 7.8 The concentration of total CO_2 in seawater is determined by the solubility of the gas and its concentration in the atmosphere. CO_2 is very soluble because it reacts with water to form HCO_3^- (and CO_3^{2-}). From reaction 6.2, it follows that an increase in atmospheric concentrations of CO_2 tends to increase the total CO_2 concentration in seawater.

Question 7.9 The 'percentage in solution' of sulphate in seawater is much too high to be accounted for by rock weathering—even though some is contributed from decomposition of sulphide and sulphate minerals. An obvious extra source is the DMS formed by plankton (Sections 6.1.3 and 7.2), along with other sulphur-rich gases, e.g. H_2S. Industry is another source you may have thought of: 65 million tonnes of sulphur added to the atmosphere annually (mainly as SO_2), compared with about 80 million tonnes from all natural sources combined. DMS contributes about one-quarter of that, according to Table 6.2: 40 million tonnes of DMS contain about 20 million tonnes of sulphur.

ACKNOWLEDGEMENTS

The Course Team wishes to thank the following: Dr. Martin Angel and Dr. Derek Pilgrim, the external assessors; Mr. Mike Hosken and Mrs. Mary Llewellyn for advice and comment on the whole Volume; Dr. Derek Pilgrim for major contributions to Chapter 5, and Dr. Ralph Rayner for help with Section 5.2.1. Dr. Malcolm Howe and Dr. Chris Vincent also provided helpful advice on content and level.

The structure and content of this Volume and of the Series as a whole owes much to our experience of producing and presenting the first Open University Course in Oceanography (S334), from 1976 to 1987. We are grateful to those people who prepared and maintained that Course, to the students and tutors who provided valuable feedback and advice and to Unesco for supporting its use overseas.

Grateful acknowledgement is also made to the following for material used in this Volume:

Figures 1.4 and 1.7 British Antarctic Survey; *Figures 1.5, 1.6 and 2.3* NASA; *Figure 2.4(a)* R. A. Horne (1969) *Marine Chemistry*, Wiley; *Figures 2.5 and 2.12* A. N. Strahler (1963) *Earth Sciences*, Harper and Row; *Figures 2.7, 4.5(a), 4.6 and Table 4.1* G. L. Pickard and W. J. Emery, *Descriptive Physical Oceanography—An Introduction*, 4th edn, Pergamon Press; *Figures 2.6(a), 2.8, 3.3(a), 4.1, 5.7, 6.7 and Table 1.1* H. U. Sverdrup *et al.* (1942) *The Oceans*, Prentice-Hall; *Figures 2.10 and 2.11* G. Haber (1977) in *New Scientist*, **73**, New Science Publications; *Figure 3.2(a)* Centro de Caridade Nossa Senhora do Perpetuo Socorro, Oporto; *Figure 3.4(a)* R. V. Tait (1968) *Elements of Marine Ecology*, Butterworths; *Figure 4.2* Estate of Mrs. J. C. Robinson; *Figures 4.9–4.11* M. C. Gregg (1973) in *Scientific American*, **228**, W. H. Freeman; *Figure 4.15* R. H. Stewart (1985) *Methods of Satellite Oceanography*, Scripps Institution of Oceanography/EROS Data Center; *Figure 4.16* K. K. Turekian (1976) *Oceans*, 2nd edn, Prentice-Hall; *Figure 5.1* R. S. Dietz (1969) in *Readings in the Earth Sciences*, Volume 2, W. H. Freeman; *Figure 5.4 (part)* N. B. Marshall (1954) *Aspects of Deep-Sea Biology*, Hutchinson; *Figure 5.6* Photo: Steve Johnson, Media Services, Plymouth Polytechnic; *Figure 5.9* US Govt. Printing Office; *Figures 5.10(a) and (b) and 5.11* D. G. Tucker and B. K. Gazey (1966) *Applied Underwater Acoustics*, Pergamon; *Figure 5.13* Courtesy Admiralty Research Establishment, Portland; *Figures 5.14 and 5.17* J. Northrop and J. G. Colborn (1977) in *Journal of Geophysical Research,* **79**, American Geophysical Union; *Figure 5.15(b)* D. Behringer *et al.* (1982) in *Nature*, **299**, Macmillan Journals; *Figure 5.16* R. C. Spindel (1982) in *Oceanus*, **25**, Woods Hole Oceanographic Institution; *Table 6.1* T. A. Davies and D. S. Gorsline; *Figures 6.2(b) and 6.3(a)* R. W. Jordan and M. Smithers; *Figure 6.3(b)* D. G. Jenkins; *Figure 6.3(c)* J. D. Milliman (1974) *Marine Carbonates*, Springer-Verlag; *Figure 6.5* W. S. Broecker and T. S. Peng (1982) *Tracers in the Sea*, Lamont-Doherty Observatory; *Figure 6.6* H. Friedrich (1969) *Marine Biology*, Sidgwick and Jackson; *Figures 6.11 and 7.1* M. Whitfield (1982) in *New Scientist*, 1 April; *Figure 6.14* W. S. Broecker (1974) *Chemical Oceanography*, Harcourt, Brace, Jovanovich, Inc.; *Figure 6.16* F. Sclater *et al.* (1976) in *Earth and Planetary Science Letters*, **31** Elsevier; *Figure 7.2* J. M. Barnola *et al.* (1987) in *Nature*, **329**, Macmillan Journals; *Figure 7.4(b)* A. Neftel (1985) in *Nature*, **315**, Macmillan Journals.

INDEX

Note: page numbers in italics refer to illustrations; in bold to tables